Designing Schools

Learning Systems that Cultivate Thinking in the Age of AI

LaShae Lee, PhD
Illustrations by Christine Lee

Human Learning by Design™

Human Learning by Design™

A series exploring learning, leadership, and human development
in an age shaped by artificial intelligence.

AI and the Art of Teaching
AI and the Art of Teaching Workbook
Designing Schools

Forthcoming titles in the Human Learning by Design™ series include:
Designing Confidence
Designing Inquiry
Designing Judgment
Designing Learning Environments
Designing Professional Identity
Designing Support

The developmental Human Learning by Design™ series include:
Becoming Human: Early Childhood
Becoming Human: Middle Childhood
Becoming Human: Adolescence
Becoming Human: Early Adulthood
Becoming Human: Adulthood
Becoming Human: Later Adulthood

Related titles in the Human Learning by Design™ collection:
Raising Humans by Design
Courage Under Algorithm for Teens
Judgment Over Speed: Leadership in an AI Workplace

ISBN: 978-1-972397-04-6 (paperback)
ISBN: 978-1-972397-05-3 (e-book)

Illustrations and cover design by Christine Lee

First Edition, 2026

Published and printed in the United States of America

CONTENTS

Designing Schools

DEDICATION

For instructional coaches and district leaders who believe schools can be intentionally designed to cultivate thinking. With appreciation for Melissa Robertson, whose leadership shaped my understanding of instructional systems, and for David M. Zirkle, Ed.D., and Jeanne Willard, Ed.D., whose work has long exemplified the kind of leadership I admire.

PROLOGUE

Artificial intelligence is acting as a pressure test for instructional systems. It reveals when efficiency outpaces meaning, when schools adopt tools without philosophical clarity, and when systems accelerate practices that were already misaligned with learning.

Learning research has long warned that visible performance is not the same as understanding. Artificial intelligence does not create weaknesses in instructional systems; it exposes them.

For decades, schools have often equated visible performance with understanding. Yet research in the learning sciences shows that "to develop competence in an area of inquiry, students must... understand facts and ideas in the context of a conceptual framework" (Bransford et al., 2000, p. 16). Learning itself is not purely informational. As Darling-Hammond and colleagues explain, "learning is shaped by the interaction of cognitive, social, and emotional processes" (2020, p. 99). Engagement also matters. According to Chi and Wylie, "different modes of engagement produce different learning outcomes" (2014, p. 220).

These findings underscore a critical distinction between the appearance of learning and the processes that produce it. Artificial intelligence does not disrupt learning itself. It reveals when systems have mistaken performance for understanding.

PART I – INSTRUCTIONAL DESIGN

Chapter 1: From Delivery Model to Design Model

District leaders may assume instructional improvement requires better programs, tighter pacing guides, or stronger compliance with standards. Yet across classrooms, the same pattern appears: some lessons ignite thinking while others produce quiet completion. The difference is rarely the curriculum itself. It is the architecture of the learning experience.

Learning does not occur in isolation. Research on human development demonstrates that "the brain is biologically wired to learn from social contexts and relationships" (Darling-Hammond et al., 2020, p. 99). Classrooms that prioritize collaboration, discussion, and shared inquiry align more closely with how humans naturally learn.

For more than a century, most school systems have operated on a delivery model of instruction: curriculum is adopted, standards are unpacked, pacing guides are distributed, and teachers deliver content. Students' complete assignments, and assessments measure the resulting output.

At scale, this model appears efficient. It promises consistency, predictability, and coverage. But it was designed in an era when information was scarce. Today we operate in an environment where information is abundant, and answers can be generated instantly. The central question for districts is no longer how to deliver content efficiently, but what kinds of thinkers our systems are designed to produce.

This chapter reframes district-level instruction not as distribution, but as architecture. Not as pacing. As formation. The distinction is simple but consequential: delivery systems move information; design systems engineer the conditions under which understanding forms.

Teaching as Cognitive Design, Not Content Transmission

Teaching has never been primarily about moving information from expert to novice. That framing collapses under AI immediately.

AI can transmit information faster than any human system ever could. What AI cannot do is design the cognitive conditions under which understanding forms. At the system level, this distinction matters.

A delivery model asks:

- Are teachers covering standards?
- Are pacing guides aligned?
- Are assessments administered on time?
- Are benchmark scores increasing?

A design model asks:

- Are students constructing conceptual understanding?
- Are they developing adaptive expertise?
- Are they practicing judgment under ambiguity?
- Are they building cognitive stamina?
- Are they transferring knowledge across contexts?

Transmission moves content. Design shapes cognition. And cognition is what endures. As Hatano and Inagaki explain, adaptive experts are distinguished by their ability to "modify knowledge when facing novel situations" (1986, p. 266)

When districts treat curriculum as a set of topics to be delivered, they inadvertently prioritize exposure over ownership. When districts treat instruction as design, they begin to examine the architecture of learning itself:

- Where does cognitive tension occur?

- Where is struggle structured and supported?
- Where is reasoning made visible?
- Where is revision normalized?
- Where is defense required?
- Where is transfer practiced?

Consider two classrooms studying ecosystems. In one classroom, students read a chapter, complete vocabulary questions, and take a quiz on food chains. In another classroom, students are asked to predict what will happen if a predator disappears from a simulated ecosystem. They model energy flow, defend predictions, and revise explanations as conditions change. Both classrooms cover the same topic. Only one is designed for thinking. The difference is not teacher enthusiasm or curriculum choice. It is design.

Instruction is not an event. It is an engineered experience.

The Anatomy of a Learning Experience

Learning environments shape cognitive risk. Students rarely attempt difficult thinking if participation threatens identity or belonging. As Hammond explains, "culture, it turns out, does matter for learning" (2015, p. 49). When belonging is uncertain, cognitive risk becomes socially dangerous, and students often protect identity by avoiding intellectual exposure. When classrooms ignore the cultural and relational context of learners, cognitive challenge can feel like exposure rather than opportunity. Designing schools therefore requires more than curriculum alignment; it requires intentional attention to the relational architecture that determines whether students interpret challenge as growth or threat.

If we treat instruction as architecture, then every learning experience has structural components. At the system level, coherence depends on understanding these components explicitly. When instruction is treated as architecture, leaders must be able to see its structural components. A coherent learning experience typically includes five interdependent elements. A learning experience includes:

1. **Entry Conditions:** Emotional tone. Clarity of purpose. Activation of prior knowledge. Signal of belonging.

2. **Cognitive Tension:** A question, problem, or contradiction. A demand that exceeds automatic recall. A reason to think.

3. **Productive Struggle:** Time to wrestle. Strategic support. Visible thinking. Peer reasoning

4. **Feedback and Adjustment:** Clarification. Correction. Extension. Revision.

5. **Transfer and Reflection:** Application in a new context. Articulation of reasoning. Defense of thinking. Metacognitive awareness.

In delivery systems, only the content and output tend to be monitored. In design systems, the architecture itself becomes visible.

District leaders can ask: Are these components systematically present across classrooms – or do they depend entirely on individual teacher expertise? If instructional architecture lives only in exceptional classrooms, then coherence does not exist.

Performance vs. Formation

Human learning is never purely cognitive. Emotion and meaning shape what students attend to, remember, and attempt. Neuroscience increasingly confirms what teachers observe daily: "emotion and cognition are supported by overlapping neural systems" (Immordino-Yang, 2016, p. 18). When learning environments ignore emotional context, cognitive engagement weakens. Designing schools for thinking therefore requires designing environments where students feel psychologically safe enough to sustain intellectual effort.

Performance is measurable in the short term. Formation is visible over time. Meaningful learning requires organizing knowledge within conceptual frameworks that support transfer beyond immediate tasks (Bransford et al., 2000).

Performance asks: Did students complete the task correctly?

Formation asks: What kind of thinker is emerging?

Delivery systems overemphasize performance because they align with accountability frameworks. Benchmarks, proficiency rates, and completion metrics are easier to quantify than identity formation or epistemic growth.

But AI destabilizes performance as a reliable indicator. If polished output can be generated externally, then performance artifacts alone cannot guarantee understanding. Formation becomes the only sustainable metric. When answers can be generated instantly, the ability to think, revise, and defend reasoning becomes the only reliable signal of understanding.

Formation includes:

- Intellectual humility
- Persistence under ambiguity
- Willingness to revise
- Capacity to defend reasoning
- Ethical discernment
- Transfer across disciplines
- Disciplined originality

These cannot be mass-produced. They must be cultivated.

A district committed to design asks not only: Are scores improving? But also: Are we forming graduates who can reason when answers are abundant?

Design Principles Across Grade Levels

Design coherence does not mean identical practice. It means shared principles. These principles act as architectural commitments that guide instruction across classrooms and grade levels. Across elementary, middle, and secondary levels, coherent systems align around several core design commitments:

Principle 1: Explanation Over Speed. Students must articulate reasoning, not merely produce answers.

Principle 2: Revision as Normal. Error must be treated as data, not deficiency.

Principle 3: Structured Struggle. Challenge must be sufficient, supported, time-bound, and debriefed.

Principle 4: Inquiry as Engine. Questions must drive instruction, not simply check comprehension.

Principle 5: Transfer as Evidence. Understanding must be demonstrated in unfamiliar contexts.

Principle 6: Belonging Before Exposure. Students must feel safe enough to risk thinking publicly.

When these principles operate system-wide, grade levels vary in sophistication, not philosophy. Without shared principles, districts drift. With them, coherence becomes possible.

Design must be structural, not accidental.

Why Compliance-Driven Systems Collapse Under AI

When design principles are absent, systems may default to compliance structures. Compliance-driven systems rely on task completion, pacing fidelity, behavioral quietness, assignment submission, and coverage metrics.

These signals once served as rough proxies for learning. Under AI, they collapse.

A student can complete assignments without constructing understanding. A classroom can appear orderly without cultivating thought. A pacing guide can be followed without deep cognition occurring. In many classrooms, a large portion of routine assignments can now be completed with generative tools without requiring meaningful student reasoning. AI does not create these weaknesses. It exposes them.

Compliance-driven systems struggle under AI because they equate output with understanding, speed with intelligence, silence with engagement, and coverage with mastery.

When answers are abundant, compliance becomes hollow.

Design-based systems, however, grow stronger under AI because they:

- Require explanation.
- Require defense.
- Require adaptation.
- Require transfer.
- Require visible reasoning.

AI can generate answers. It cannot defend them under live questioning. It cannot revise meaningfully based on contextual nuance. It cannot build relational trust.

A district that continues to operate as a delivery mechanism accelerates misalignment. A district that shifts to design cultivates resilience.

District Tools

The move from delivery to design requires structured diagnostics. Below are system-level tools for leadership teams.

Instructional Coherence Audit

Purpose: Determine whether district practices reflect shared design principles or isolated excellence.

Audit Questions:

- Do grade-level teams share a common definition of understanding?
- Are explanation and defense required across content areas?
- Is revision structurally embedded in curriculum?
- Are inquiry sequences vertically aligned?
- Do observation protocols measure cognitive architecture – or only compliance and pacing?
- Are feedback systems identity-safe and growth-oriented?

Warning Indicators:

- High variability in classroom cognitive demand
- Overemphasis on pacing compliance
- Limited student-to-student reasoning
- Assessment focused primarily on correctness

Coherence is not uniformity. It is alignment around design commitments.

Delivery-to-Design Diagnostic

Purpose: Surface where district systems remain delivery-driven and where they are shifting toward visible instructional design.

Compare Current Practice Across Four Domains

Domain	Delivery Model Signal	Design Model Signal
Curriculum	Coverage of topics	Sequencing of cognitive tensions
Assessment	Correctness of output	Defense & transfer of reasoning
Instruction	Teacher explanation dominant	Student reasoning visible
Observation	Pacing and engagement	Cognitive architecture and struggle

Leadership teams can identify where they are operating as delivery and where they have shifted to design. The goal is not rhetorical agreement. It is structural evidence.

Curriculum Signal Analysis Tool

Purpose: Determine whether a curriculum unit functions as delivery (task completion) or design (reasoning, revision, and transfer).

Every curriculum sends signals about what counts. Analyze one unit and ask:

- Does this unit reward speed or explanation?
- Does it normalize revision?
- Does it require defense?
- Does it demand transfer?
- Does it structure productive struggle?
- Does it protect belonging during exposure?

If AI can complete 80% of the tasks without requiring student reasoning, the unit is delivery driven. Design requires friction.

PLC Protocol

Purpose: Reveal whether a curriculum unit signals task completion or requires reasoning, revision, and transfer.

In a Professional Learning Community (PLC) session, select one upcoming unit, then compare it through both lenses.

Step 1: Analyze Through the Delivery Lens

- What content is covered?
- What assignments are completed?
- What assessments measure mastery?
- What pacing constraints exist?

Step 2: Analyze Through the Design Lens

- What cognitive tensions are engineered?
- Where is reasoning required?
- Where is revision expected?
- Where is transfer demonstrated?
- What kind of intellectual posture does this unit cultivate?

Step 3: After analysis, ask two central questions:

1. What does this unit reward: speed, compliance, and completion, or reasoning, persistence, and defense?
2. What kind of thinker does it produce: a reproducer of information or an adaptive, judgment-capable learner?

Document the differences. Effective design conversations move from abstract philosophy to unit-level architecture.

The Learning Architecture Model

District redesign in the age of artificial intelligence requires coherence across five system layers:

1. **Instructional Design:** How learning experiences are structured to cultivate thinking.

2. **Professional Expertise:** How teachers and coaches develop adaptive instructional judgment.

3. **Ethical AI Integration:** How technology is governed so that it supports – rather than replaces – cognition.

4. **System Alignment:** How curriculum, assessment, and evaluation reinforce the same intellectual goals.

5. **Leadership Implementation:** How districts build capacity for sustained instructional redesign.

The structure of this book intentionally mirrors the five-layer learning architecture model introduced above. Each section examines one layer of system redesign so that district leaders can move from philosophy to implementation in a coherent progression.

Part I – Learning Architecture establishes the foundational shift from delivery models of instruction to design models of learning. It examines instructional architecture, backward design in an AI era, and cognitive load considerations that shape how learning experiences cultivate thinking.

Part II – Professional Expertise explores how districts build adaptive instructional judgment among educators. These chapters focus on coaching for design thinking, developing teacher expertise, and creating professional learning structures that support continuous instructional redesign.

Part III – Ethical AI Integration addresses the governance and moral considerations of artificial intelligence in schools. Rather than treating AI as a technical tool, these chapters examine how districts ensure technology supports human cognition, protects student dignity, and aligns with educational values.

Part IV – System Alignment examines how curriculum, assessment, grading, and accountability systems must reinforce the same intellectual goals. Without alignment across these structures, even strong instructional design cannot scale across a district.

Part V – Implementation translates these ideas into leadership action. This section provides district leaders with practical structures – including a 90-day redesign plan – to move from conceptual agreement to sustained system change.

Together, these five sections form a coherent leadership framework. District redesign in the age of artificial intelligence cannot occur through isolated initiatives. It requires alignment across instructional architecture, professional expertise, ethical technology governance, system structures, and leadership implementation. This book follows that progression deliberately so leaders can move from understanding the challenge to designing systems that endure.

When these five layers align, schools move beyond content delivery and begin to cultivate graduates capable of reasoning, adapting, and exercising judgment in an AI-rich world.

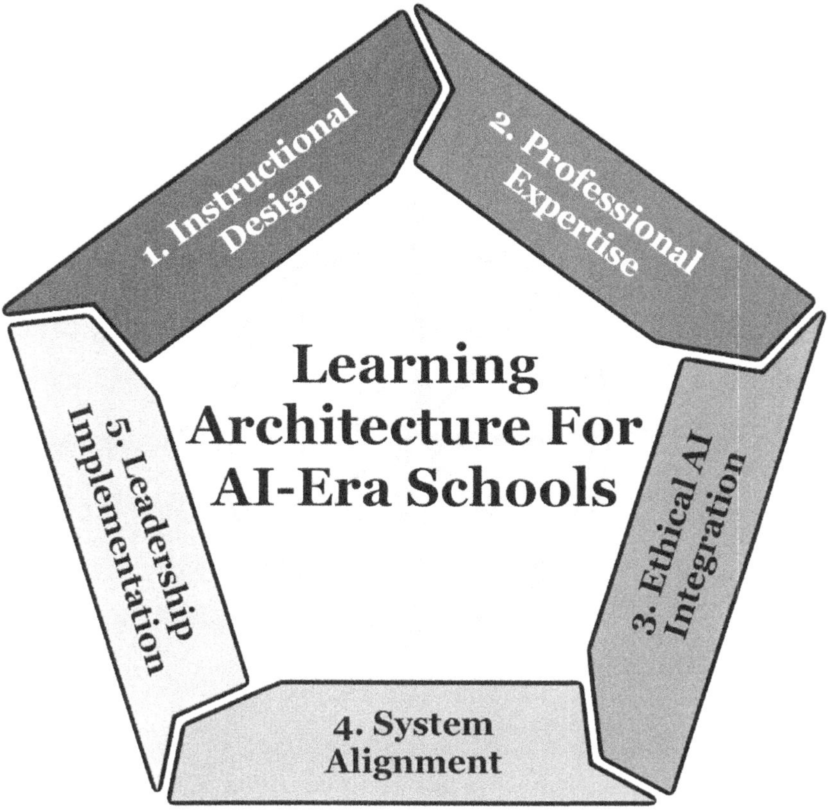

Figure 1: The Designing Schools Framework. Five interdependent layers of district redesign in an AI-rich learning system. Layers are (1) Instructional design, (2) professional expertise, (3) ethical AI integration, (4) system alignment, and (5) leadership implementation.

Leadership Reflection

Artificial intelligence has not disrupted teaching. It has disrupted illusions. Illusions that output equals understanding. Illusions that coverage equals learning. Illusions that compliance equals engagement.

Districts now face a defining choice: accelerate delivery or redesign learning architecture. Delivery systems move information efficiently. Design systems cultivate thinkers.

Sustainable redesign cannot occur in a single classroom or through a single initiative. It requires coherence across the layers that shape learning – how instruction is designed, how professional expertise develops, how technology is governed, how systems align, and how leadership sustains implementation.

The systems that endure will not be the ones that move content fastest, but the ones that cultivate judgment most intentionally.

The work of redesign begins with design clarity. The chapters that follow explore how that clarity becomes system architecture. The shift from delivery to design is expressed through concrete leadership actions.

Superintendent Briefing: From Delivery Systems to Learning Architecture

Artificial intelligence is not simply a technological development; it is a systems test for education. Tools that generate answers instantly expose the underlying design assumptions of instructional systems. When learning environments are built around coverage, completion, and correct answers, AI can perform many of those tasks faster than students. The result is not merely a technological challenge but a structural one.

District leaders therefore face a choice. Systems can accelerate information delivery, or they can redesign learning architecture to cultivate reasoning, judgment, and intellectual independence. The latter requires coherence across the five system layers introduced in this chapter: instructional design, professional expertise, ethical AI integration, system alignment, and leadership implementation.

For superintendents, this shift has practical implications. Curriculum adoption must is evaluated for the kinds of thinking tasks it requires. Professional learning evolves beyond strategy workshops toward instructional design expertise. Technology governance ensures that artificial intelligence supports cognition rather than replacing it. Assessment and grading policies reinforce explanation, transfer, and revision rather than speed and completion.

The central leadership question becomes straightforward: Do our systems reward the production of answers, or the development of thinkers?

District redesign begins when leadership decisions consistently reinforce the latter.

System Leader Playbook: From Delivery to Design

Shifting from a delivery model of instruction to a design model does not occur through slogans or isolated initiatives. It occurs when district leaders adjust the structures that shape classroom practice – how instruction is observed, how tasks are designed, and what kinds of thinking systems signal as valuable. Leadership decisions influence whether classrooms reward rapid completion or sustained reasoning.

The following actions provide an initial entry point for system leaders seeking to translate design principles into district practice.

Leaders can begin shifting from delivery by design by:

- Auditing cognitive architecture in classrooms
- Protecting productive struggle in instruction
- Revising observation tools to measure thinking
- Clarifying that explanation matters more than completion
- Redesigning tasks that AI can complete without reasoning

These actions help translate the shift from delivery to design into visible leadership practice. When system structures reinforce reasoning, revision, and transfer, classrooms increasingly reflect the architecture required for durable understanding.

Chapter 2: Backward Design in an AI Era

In Chapter 1, district redesign was described as requiring coherence across five system layers: instructional design, professional expertise, ethical AI integration, system alignment, and leadership implementation. The first of these layers – instructional design – shapes the daily learning experiences students encounter in classrooms.

Instructional design determines whether learning environments reward completion or cultivate reasoning – whether students memorize answers or develop the capacity to transfer understanding across unfamiliar situations.

In an era where artificial intelligence can generate explanations, summaries, and solutions instantly, the architecture of instructional design becomes even more consequential. When learning experiences are built around information recall, AI can easily replace student effort. When learning experiences require explanation, interpretation, and judgment, human thinking remains central.

This chapter therefore examines how backward design must evolve in an AI era so that instructional architecture continues to cultivate durable understanding.

Instructional design strongly influences learning outcomes. A large meta-analysis of STEM education found that "interactive engagement methods are more effective than traditional lectures for promoting conceptual understanding" (Freeman et al., 2014, p. 8410). Learning environments that require explanation, discussion, and problem-solving produce deeper understanding than environments dominated by passive reception.

Backward design transformed curriculum planning by asking educators to begin with the end in mind. Instead of starting with activities or textbook chapters, leaders began with enduring understandings and essential questions, then aligned assessment and instruction accordingly.

That shift was profound. But AI changes the landscape again. If students can generate polished responses to most traditional prompts instantly, then we must ask a more unsettling question:

Backward from what?

In an AI-saturated world, the "end" cannot simply be correct answers or coherent essays. Those are now reproducible outputs. Backward design must now begin with something AI cannot independently cultivate: Human judgment. This chapter reimagines backward design at the system level for an answer-abundant era.

Enduring Understandings vs. Answerable Prompts

Traditional backward design distinguished between topics (what we cover) and enduring understandings (what lasts beyond the lesson).

But in practice, many units still center on answerable prompts: Define photosynthesis. Explain the causes of the Civil War. Solve for x. Summarize the article. These tasks once required cognition. Today, many can be generated fluently by AI tools within seconds.

The distinction districts now clarify is this: enduring understandings shape thinking, while answerable prompts produce artifacts. An enduring understanding might be:

- Energy transforms but is never destroyed.
- Power structures influence historical narratives.
- Mathematical models approximate reality under constraints.
- Claims require evidence and counterargument.

An answerable prompt asks for reproduction. An enduring understanding demands interpretation.

District leaders can ask whether curriculum is structured around the coverage of answers or the cultivation of insight. Backward design in an AI era requires clarity about what cannot be outsourced.

AI-Resistant Transfer Goals

Transfer has always been the gold standard of understanding. In an AI-rich environment, however, transfer goals benefit from being intentionally designed to resist automation.

An AI-resistant transfer goal typically asks students to:

- Apply knowledge in a novel, context-bound scenario
- Justify reasoning under constraints
- Weigh competing values or interpretations
- Defend decisions publicly
- Revise based on feedback

For example:

Low-Resistance Goal: "Write an argumentative essay about renewable energy." AI can complete this.

High-Resistance Transfer Goal: "Advise a city council with competing economic and environmental interests on which renewable energy investment to pursue. Defend your recommendation under live questioning and revise your proposal based on counterarguments."

This goal requires contextual judgment, ethical positioning, adaptation, and defense of reasoning. AI can draft language, but it cannot own the reasoning under scrutiny.

Transfer tasks therefore include ambiguity, constraint, defense, and adaptation. When a task can be completed in isolation without contextual negotiation, it remains vulnerable to automation.

Designing for Judgment, Not Recall

Recall measures memory. Judgment measures reasoning under uncertainty.

Backward design now anchors around questions such as:

- What decisions will graduates need to make in the real world?
- Under what ambiguity must they operate?
- What values will they need to weigh?
- What tradeoffs must they consider?
- What misconceptions must they navigate?

Judgment involves interpreting incomplete data, reconciling contradiction, anticipating consequences, exercising ethical discernment, and engaging in strategic reasoning. Recall is important but insufficient. When districts over-index on recall-based standards, they unintentionally design for obsolescence in an AI era.

Designing for judgment includes learning experiences where:

- Students must explain why.
- Students must defend how.
- Students must adapt when conditions change.
- Students must revise publicly.

Judgment cannot be measured through multiple-choice accuracy alone. It requires performance under tension.

What "Understanding" Now Requires

Understanding once implied correct explanation, accurate solution, and coherent summary. In an AI-rich environment, understanding implies more. Understanding requires:

1. **Conceptual Coherence:** Students can explain underlying structures, not just outcomes.

2. **Adaptive Application:** Students can transfer knowledge into unfamiliar conditions.

3. **Metacognitive Awareness:** Students can articulate how they know and why they chose a strategy.

4. **Defense Under Scrutiny:** Students can justify reasoning in dialogue.

5. **Ethical Positioning:** Students can recognize implications beyond correctness.

When understanding does not include transfer and defense, it remains vulnerable to automation. Backward design therefore must evolve accordingly.

The Human Capability Backward Design Model

Backward design in an AI-rich environment begins with a different question: What forms of human capability should schooling cultivate? When answers, summaries, and explanations can be generated instantly, curriculum design cannot begin with the production of answers alone. It begins with the development of judgment.

Rather than designing backward from correct responses, districts design backward from the capabilities students need when conditions are ambiguous, information is abundant, and decisions carry consequences.

The Human Capability Backward Design Model includes three stages:

1. **Human Capability Outcomes:** Define the forms of reasoning, judgment, and transfer students must develop.

2. **Transfer & Judgment Evidence:** Determine what forms of performance demonstrate ownership of reasoning.

3. **Cognitive Tension Architecture;** Design learning experiences that require explanation, revision, and adaptation.

Together these stages reframe backward design for a world where answers are abundant but judgment remains scarce.

Stage 1 - Human Capabilities Outcomes

Traditional Stage 1: Identify desired results.
AI-Era Stage 1: Identify the human capabilities learning should cultivate.

District curriculum teams define outcomes such as intellectual humility, analytical reasoning, ethical discernment, transfer across domains, persistence under cognitive load, structured disagreement, and creative constraint navigation.

Instead of asking: What should students know?

Leaders ask: What should students be able to do that AI cannot independently do for them?

Capabilities can then be documented vertically across grade bands so that expectations deepen over time. This progression becomes the district's design spine, clarifying how reasoning develops from elementary through secondary levels.

Stage 2 — Transfer & Judgment Evidence

Traditional Stage 2: Determine acceptable evidence.
AI-Era Stage 2: Identify evidence of transfer and judgment.

Evidence of understanding now includes performances such as live defense, oral reasoning, revision cycles, context-bound decision making, counterargument response, and cross-disciplinary application.

Assessment design therefore asks:

- Can students adapt when variables change?
- Can they defend reasoning without prewritten script?
- Can they revise meaningfully?

Evidence shifts from product to process. The central question becomes not Did students submit it? but Can they own the reasoning behind it?

Stage 3 — Cognitive Tension Architecture

Traditional Stage 3: Plan learning experiences.
AI-Era Stage 3: Design cognitive tension intentionally.

Units incorporate entry ambiguity, structured struggle, misconception exposure, counterfactual thinking, defense moments, and transfer rehearsal.

Cognitive tension does not occur accidentally; it is engineered. District curriculum teams therefore examine questions such as:

- Where does thinking stretch?
- Where does discomfort occur safely?
- Where must students wrestle before resolution?

When tension is absent, learning becomes procedural. When tension is unsupported, students disengage. Effective design calibrates challenge and safety so that intellectual effort becomes sustainable.

District Alignment Tools

Translating instructional philosophy into classroom practice requires more than shared language about learning goals. District teams benefit from practical tools that help examine whether standards, tasks, and assessments cultivate the human capabilities schools intend to develop in an AI-rich environment.

The tools below help curriculum teams and instructional leaders analyze standards for reasoning demand, redesign tasks to require transfer and judgment, and ensure that learning experiences make student thinking visible. By mapping standards to human capabilities and strengthening transfer-oriented design, districts can align instructional planning with the broader learning architecture described throughout this book.

AI-Resilient Standards Mapping Guide

Standards are not obsolete, but their interpretation evolves in an AI-rich environment. When answers can be generated instantly, the central question becomes whether standards cultivate reasoning or simply organize information.

District teams can map each priority standard using the following diagnostic criteria.

1. Is this standard recall-dominant or judgment-dominant?

Some standards primarily ask students to reproduce information. Others require interpretation, evaluation, and reasoning.

- Recall-dominant standards require reinforcement through explanation and transfer.
- Judgment-dominant standards require defense and ambiguity.

2. What human capability does this standard build?

For each standard, identify the primary capability the learning experience develops, such as:

- Analytical reasoning.
- Ethical evaluation.
- Argumentation.
- Modeling.
- Interpretation.
- Creative synthesis.

When a standard cannot be clearly linked to a capability, the instructional design around it likely benefits from redesign.

Example AI-Resilient Standards Mapping:

Priority Standard	Recall or Judgment Dominant	Human Capability Developed	AI Vulnerability	Assessment Redesign Needed
Analyze energy transfer in ecosystems	Judgment	Analytical reasoning	Moderate	Add defense and transfer scenario
Explain causes of the American Revolution	Recall	Interpretation	High	Add contextual decision scenario

3. Can AI complete the task aligned to this standard without student reasoning?

District teams examine whether the current assessment conditions allow AI tools to generate acceptable responses without student ownership of reasoning.

When tasks are easily automated, teams adjust assessment conditions by introducing elements such as:

- Context constraints
- Defense components
- Transfer scenarios
- Revision requirements

These structures make student reasoning visible rather than relying solely on finished products.

4. Where does this standard vertically connect?

Standards become more powerful when cognitive demand develops progressively across grade bands.

District teams map how expectations deepen:

Elementary → Explanation
Middle → Application under variation
High School → Defense under counterargument
Postsecondary → Ethical and interdisciplinary synthesis

Coherence emerges when standards gradually cultivate judgment rather than repeatedly rehearsing recall.

Transfer Design Checklist

Before finalizing a unit, teams can examine whether the learning experience requires students to exercise reasoning rather than simply retrieve information.

Key questions include:

- Does this task require reasoning beyond information retrieval?
- Does it introduce contextual constraints?
- Does it require defense of reasoning?
- Does it allow revision after critique?
- Does it require application in a new context?

When a task can be completed without contextual judgment, it remains vulnerable to automation. In these cases, teams may strengthen the design by incorporating transfer conditions, defense moments, or revision cycles.

Instructional design represents the first layer of district learning architecture. When backward design is grounded in human capability and transfer, the remaining layers – professional expertise, ethical AI governance, and system alignment – can develop with greater coherence.

Leadership Reflection

Design cannot begin with content coverage, benchmark performance, or essay production alone. Instead, curriculum design begins with questions such as:

- Human capability
- Transfer under ambiguity
- Ethical judgment
- Cognitive endurance

Artificial intelligence does not eliminate the need for rigorous curriculum. It clarifies what rigor truly means. Systems designed for judgment remain resilient in an AI era. Systems designed around answer production rely increasingly on detection tools.

Backward design in an AI-rich environment begins with a simple but destabilizing question: When answers are abundant, what remains unmistakably human?

That is the outcome worth designing backward from.

Superintendent Briefing: Redesigning Backward Design in an AI Era

Backward design has long provided a useful structure for curriculum planning. Educators begin by identifying desired results, determine evidence of learning, and then design instructional experiences. In an environment where artificial intelligence can generate explanations, summaries, and solutions instantly, however, the definition of "desired results" deserves renewed scrutiny.

When desired results are framed primarily as correct responses or completed products, AI tools can often produce those outcomes independently. In these cases, assessment reflects technological efficiency more than student understanding. District leaders therefore face an emerging design challenge: ensuring that curriculum defines outcomes in terms of human capability rather than answer production.

Desired results increasingly center on reasoning, interpretation, explanation, ethical judgment, and the ability to apply knowledge in unfamiliar contexts. Evidence of learning similarly shifts toward transfer tasks that invite students to defend reasoning, adapt knowledge, and revise their thinking when confronted with new information.

This shift influences several district systems. Curriculum adoption benefits from examining whether units require explanation and transfer or primarily reward answer retrieval. Assessment policies increasingly emphasize reasoning over completion. Professional learning supports educators in designing tasks that reveal student ownership of thinking.

Backward design remains essential – but its endpoint has evolved. In an AI-rich environment, the goal of curriculum extends beyond correct answers toward the development of judgment that cannot be automated.

System Leader Playbook: Redesigning Backward Design

District leaders can strengthen backward design in an AI-rich environment through several practical actions:

1. **Audit Units for Answerable Prompts.** Identify assignments AI can complete independently and redesign them to require reasoning and defense.
2. **Redefine Desired Results Around Human Capabilities.** Clarify the forms of thinking graduates must demonstrate, including transfer, judgment, and ethical reasoning.
3. **Convert Assessments to Transfer Tasks.** Ensure assessments require contextual decision-making, explanation, defense, and revision.
4. **Require Evidence of Ownership.** Incorporate oral reasoning, live questioning, revision cycles, and peer critique.
5. **Align PLC Planning with Capability Development.** Shift PLC conversations from "What content will we cover?" to "What thinking will this unit cultivate?"

When these principles guide curriculum planning across classrooms, backward design begins to function as system architecture rather than isolated lesson planning.

Chapter 3: Cognitive Load and Instructional Clarity

Designing Attention and Friction in an AI-Rich System

Chapter 1 introduced the Learning Architecture Model, which describes five layers of district redesign in the age of artificial intelligence. The first layer – instructional design – shapes the daily learning experiences that cultivate reasoning, transfer, and intellectual independence. Chapter 2 examined how backward design must evolve when artificial intelligence can generate answers instantly. Instead of designing backward from correct responses, curriculum must now begin with human capabilities such as explanation, interpretation, and judgment.

This chapter completes the first layer of the system model by examining the cognitive architecture of learning experiences themselves – how tasks, questioning, feedback, and revision structures shape the thinking students must do. Instructional design ultimately determines whether students merely produce answers or develop the capacity to reason in unfamiliar situations.

If Chapter 1 reframed teaching as architecture and Chapter 2 redesigned backward planning around human capability, this chapter turns to something even more fragile: attention.

Attention functions as the entry point to learning. Posner and Petersen describe attention as "the gateway to perception, memory, and learning" (1990, p. 25). When learning environments are saturated with distractions or fragmented tasks, sustained cognitive engagement becomes difficult. Designing schools therefore requires careful attention to how time, technology, and space shape students' ability to focus.

Learning that feels easy in the moment often disappears quickly. Cognitive psychology demonstrates that difficulty, when carefully designed, strengthens learning. Bjork and Bjork describe these challenges as "desirable difficulties" – conditions that may slow performance initially but "enhance long-term retention and transfer" (2011, p. 56). Systems designed primarily for efficiency can unintentionally remove the productive struggle necessary for durable understanding.

In an AI-rich environment, information is abundant and explanations can be generated instantly. But cognition remains constrained: attention is limited, working memory is finite, and processing capacity is narrow. Systems that ignore these constraints risk accelerating output while quietly weakening understanding.

This chapter grounds district design decisions in two complementary frameworks:

- Sweller's Cognitive Load Theory
- The Attention Architecture Model

Together, they offer leaders a disciplined way to evaluate when AI supports thinking – and when it quietly replaces it.

The Cognitive Load Foundation

John Sweller's Cognitive Load Theory clarified a fundamental truth: working memory is limited. As Sweller and colleagues explain, "working memory is severely limited in both capacity and duration when dealing with novel information" (Sweller, Ayres, & Kalyuga, 2011, p. 4).

This constraint carries direct implications for instructional design. Learning depends on how learning experiences manage cognitive load across three types:

1. **Intrinsic Load:** The inherent complexity of the material itself. Some concepts are naturally demanding.
2. **Extraneous Load:** Cognitive burden created by poor design – unclear instructions, unnecessary tasks, distracting visuals, split attention.
3. **Germane Load:** The productive mental effort devoted to constructing understanding.

In traditional classrooms, leaders often focused primarily on reducing extraneous load created by unclear explanations or inefficient materials. In AI-rich classrooms, the equation shifts. Artificial intelligence can reduce cognitive load – and it can also remove the very struggle that strengthens learning.

District leaders therefore face a new design distinction: identifying when load reduction supports cognition and when it bypasses it.

Intrinsic vs. Extraneous Load in AI-Rich Environments

Artificial intelligence introduces a paradox. It can remove unnecessary friction—but it can also remove the productive struggle that supports learning.

Productive uses of AI often reduce extraneous load, such as:

- Summarizing a dense article before close reading
- Translating academic vocabulary for multilingual learners
- Generating practice problems to build fluency
- Automating formatting to focus on reasoning

In these situations, AI reduces extraneous cognitive burden and frees working memory for deeper thinking.

Riskier uses of AI eliminate germane load, the effort required to construct understanding. Examples include:

- Generating the full essay instead of structuring the argument
- Solving the complex problem instead of modeling strategy
- Writing code instead of debugging logic
- Producing analysis without wrestling with the data

In these cases, AI removes the cognitive work that builds schema and supports transfer. The result is polished output but limited ownership of reasoning.

For system leaders, the central design question becomes straightforward: Are we reducing distraction – or reducing thinking?

When AI Reduces Load Productively

Artificial intelligence is not the enemy of cognitive science. When used intentionally, it can support cognitive efficiency. Productive uses of AI often reduce extraneous load while preserving the cognitive effort required for understanding.

Examples include:

1. **Scaffolding Novices.** Beginning learners experience overwhelming intrinsic load when encountering complex ideas for the first time. AI tools can provide simplified explanations, worked examples, and vocabulary previews that help students access the material. These supports reduce overload while preserving the conceptual demands of the task.
2. **Automating Peripheral Tasks.** When the instructional goal centers on reasoning, some tasks function primarily as extraneous load. AI can assist with formatting citations, generating practice variations, or creating rehearsal questions. Automating these peripheral tasks allows students to direct attention toward conceptual work.
3. **Increasing Feedback Speed.** AI-assisted feedback can shorten the time between attempt and revision. When feedback arrives quickly, students can refine their thinking while the reasoning process remains active, preserving germane cognitive effort.

The central design question remains: Is AI supporting schema construction – or substituting for it?

When AI Erodes Desirable Difficulty

Not all difficulty is harmful. Robert Bjork's concept of desirable difficulty reminds us that certain forms of effort strengthen learning.

Struggle that is time-bound, structured, supported, and debriefed enhances retention and transfer.

AI can erode desirable difficulty when students outsource initial attempts, when drafting becomes optional, when revision begins from a fully polished baseline, or when explanation is replaced by paraphrasing.

When the first draft is machine-generated, the cognitive work of structuring thought disappears. District leaders therefore face an important design distinction: efficiency does not necessarily produce understanding.

Cognitive growth requires:

- Retrieval
- Organization
- Synthesis
- Self-explanation

These processes are effortful by design. Removing them removes learning.

Designing Friction Intentionally

Friction is not a flaw. It is an instructional tool. Effective friction, however, is intentional, targeted, and proportionate.

Unnecessary friction – confusing instructions, unclear expectations, poorly organized materials – drains cognitive capacity. Productive friction – problem solving, defense of reasoning, synthesis – strengthens it.

District leaders can help teachers distinguish between design flaws and design features.

Friction becomes a feature when it:

- Forces retrieval
- Requires explanation
- Demands adaptation
- Invites revision
- Encourages metacognition

Without friction, cognition becomes passive. With unmanaged friction, cognition collapses. Effective system design depends on calibrating challenge so that effort strengthens learning rather than overwhelming it.

Attention Architecture in a Distracted Age

The Attention Architecture framework adds a second dimension: Cognitive load is internal. Attention drain is environmental.

Students now navigate device notifications, multiple browser tabs, AI chat windows, social media alerts, and competing digital stimuli.

Each environmental cue competes for working memory. District policy therefore cannot treat attention as incidental. Attention is not simply a character trait – it is shaped by architecture.

Leaders can examine what their systems normalize: constant device presence, silent multitasking, and unstructured browsing.

Or they can design environments that prioritize protected thinking windows, clear device expectations, and deliberate offline intervals.

Attention is a limited resource. Effective design protects it.

District Tools

System-level coherence depends on practical tools that translate cognitive science into instructional design. The following alignment structures support district teams in examining how learning environments manage cognitive load, attention, and technology.

Lesson Load Mapping Protocol

Purpose: Examine how a lesson distributes intrinsic, extraneous, and germane cognitive load before implementation.

For each lesson, map the following dimensions.

A. Intrinsic Load

- What conceptual complexity exists?
- Is this new, layered, abstract?

B. Extraneous Load

- Are instructions clear?
- Are visuals aligned with purpose?
- Is the task sequence coherent?
- Are unnecessary steps removed?

C. Germane Load

- Where must students explain?
- Where must they retrieve?
- Where must they transfer?
- Where must they revise?

Then ask: Does AI reduce extraneous load – or germane load?

Teams document their conclusions explicitly. When AI reduces germane load, the lesson design benefits from revision.

Attention Drain Audit

Purpose: Identify environmental factors that compete with sustained cognitive engagement.

Audit three domains: device environment, instructional transitions, and physical environment.

Each domain can be rated: Low Drain. Moderate Drain. High Drain.

Device Environment

- Are devices open by default?
- Is multitasking visible?
- Are AI tools structured or unbounded?

Instructional Transitions

- How often are students shifting tasks?
- Are directions chunked?
- Is wait time protected?

Physical Environment

- Visual clutter
- Notification noise
- Seating arrangement for attention

When classrooms experience high levels of attention drain, environmental adjustments often prove more effective than motivational reminders.

Device Policy Design Framework

Purpose: Align device use with cognitive goals rather than behavioral control.

District device policies often oscillate between full restriction and full access. Neither extreme reflects what cognitive science suggests about attention and learning.

A design-based framework examines questions such as:

- When do devices support intrinsic understanding?
- When do they reduce extraneous burden?
- When do they erode desirable difficulty?
- When must devices be closed to protect deep work?

Effective device policies are purpose-driven, content-sensitive, grade-sensitive, and transparent.

Example Cognitive Alignment:

- Retrieval Practice → Devices Closed
- Drafting Phase 1 → Devices Limited
- Research Phase → Devices Structured
- Defense & Dialogue → Devices Closed

Under this approach, device policy becomes a cognitive design decision rather than a behavioral rule.

Leadership Reflection

In AI-rich systems, the temptation is acceleration: faster drafting, faster grading, faster completion. Cognitive architecture, however, does not accelerate simply because tools do.

Learning depends on attention, effort, retrieval, struggle, and revision.

Artificial intelligence can reduce unnecessary friction. It cannot replace productive effort.

Districts that overlook cognitive load often weaken learning while appearing efficient. Districts that calibrate friction intentionally cultivate cognitive stamina, transfer capacity, and disciplined reasoning.

The central question for system leaders is not: How do we integrate AI everywhere?

It is: Where should thinking remain effortful?

That distinction separates acceleration from erosion. In an AI-rich environment, erosion can resemble success — until judgment is required.

Effective design protects the mind.

When instructional design consistently requires explanation, transfer, and revision, learning environments begin to cultivate reasoning rather than compliance. Yet architecture alone is insufficient.

Even thoughtfully designed systems depend on educators capable of interpreting student thinking, adapting instruction, and exercising professional judgment. The next section therefore examines the second layer of the Learning Architecture Model: professional expertise.

District redesign ultimately depends not only on stronger learning architecture, but also on educators who can design and refine that architecture over time.

Superintendent Briefing: From Delivery to Design

Artificial intelligence reveals whether instructional systems cultivate thinking or primarily organize content coverage.

District Leadership Actions

1. **Audit instructional tasks for reasoning demands.** Examine whether learning experiences require explanation, interpretation, and defense of thinking.
2. **Redesign tasks that are easily completed by AI.** Strengthen assignments so that reasoning, transfer, and judgment become central.
3. **Align walkthrough tools with cognitive architecture.** Observation systems can focus on visible thinking, productive struggle, and opportunities for explanation.
4. **Protect productive struggle in classrooms.** Learning environments can balance challenge and support so that effort strengthens understanding.
5. **Shift grading emphasis toward reasoning and transfer.** Assessment policies can signal that explanation and adaptation matter more than simple completion.

Key Question for Leadership Teams: Are our systems producing students who complete tasks – or students who can think when conditions change?

System Leader Playbook: Designing for Cognitive Clarity

District leaders influence instructional clarity not only through curriculum decisions but also through the structures that shape classroom attention, pacing, and task design. When learning environments overload working memory or fragment attention, reasoning becomes difficult regardless of teacher effort.

Leadership Moves for System Clarity:

1. **Audit instructional materials for cognitive load.** Examine whether curriculum materials present information in manageable segments and whether examples support schema development rather than overwhelm learners.

2. **Protect focused learning time.** Schedules, device expectations, announcements, and classroom routines can minimize unnecessary interruptions so students sustain cognitive engagement.

3. **Design tasks that include productive friction.** Learning experiences can invite explanation, reasoning, and revision rather than rapid completion of procedural steps.

4. **Develop educator awareness of cognitive overload.** Professional learning can help teachers distinguish between struggle caused by intrinsic complexity and confusion created by extraneous design.

5. **Evaluate technology through a cognitive lens.** Before adopting new tools, leaders can examine whether technology reduces unnecessary cognitive burden or introduces additional distraction.

Leadership Question: Are our learning environments designed for clarity and reasoning – or are they unintentionally overwhelming students' cognitive capacity?

When districts design for cognitive clarity, they create conditions where attention can be sustained and understanding can deepen.

PART II – PROFESSIONAL EXPERTISE

Chapter 4: Adaptive Expertise Over Procedural Fluency

Designing for Judgment in an Age of Procedural Outsourcing

Chapter 1 introduced the Learning Architecture Model, which describes five layers of district redesign in the age of artificial intelligence. Chapters 2 and 3 examined the first layer – instructional design – and how curriculum, cognitive load, and attention architecture shape the thinking students must do.

But architecture alone cannot cultivate adaptive thinkers.

Even the most thoughtfully designed systems depend on educators capable of interpreting student reasoning, adapting instruction in real time, and designing variation intentionally. This second layer of the system model – professional expertise – examines the human judgment required to sustain strong instructional architecture.

This chapter begins that exploration by examining the difference between routine expertise and adaptive expertise in an AI-rich world. If Part I established teaching as architecture, Part II turns to the human capacity that architecture must cultivate: Adaptive expertise.

The ultimate test of learning is not recall but transfer. As Bransford and colleagues explain, "transfer is the ability to extend what has been learned in one context to new contexts" (2000, p. 51). Schools that emphasize completion of tasks rather than application of ideas often produce students who perform well in familiar situations yet struggle to apply knowledge beyond the classroom.

Modern learning environments must cultivate more than procedural proficiency. Hatano and Inagaki describe that "adaptive expertise are able to apply their knowledge flexibly and to invent new procedures when familiar routines are insufficient" (1986, p. 266). In an era where intelligent systems can reproduce established procedures instantly, the human advantage increasingly lies in this adaptive capacity.

For decades, school systems have primarily rewarded routine expertise. Students learned procedures. Practiced repetition. Increased speed. Demonstrated accuracy. Under stable conditions, this model functioned adequately. In an AI-rich environment, it collapses. AI excels at routine expertise.

If our curriculum is built primarily around procedural fluency, we are preparing students to compete with systems designed for procedural superiority.

This chapter argues for a system-wide shift: From routine fluency to adaptive expertise.

Routine Expertise vs. Adaptive Expertise

Hatano and Inagaki distinguish "routine experts," who apply procedures efficiently in familiar situations, from "adaptive experts," who can modify their knowledge when conditions change (1986, p. 266). Routine experts perform accurately and quickly within known boundaries but often struggle when problems depart from familiar patterns.

Adaptive experts, in contrast:

- Modify strategies under new constraints
- Generate novel approaches to unfamiliar problems
- Interpret ambiguity rather than avoid it
- Reflect metacognitively on their reasoning
- Transfer knowledge across contexts

Routine expertise values speed and repetition. Adaptive expertise values flexibility and conceptual depth. Both forms of expertise remain valuable. In an era of intelligent tools, however, adaptive expertise becomes increasingly central to human work.

AI and Procedural Outsourcing

Artificial intelligence systems now perform mathematical computation, essay structuring, code generation, data summarization, language translation, and pattern recognition. They do so with speed and fluency beyond human capacity.

When students are trained primarily to replicate procedures, AI quickly becomes a shortcut. The central risk is not dishonesty – it is dependency.

When procedures are outsourced before they are conceptually understood, students may:

- Lose generative flexibility
- Bypass productive initial struggle
- Confuse polished output with understanding
- Consume reasoning rather than produce it

The central question is not whether AI exists. It is whether students can think when AI is absent.

Adaptive expertise depends on internalized schema. Without schema, transfer collapses.

Designing Novelty into Curriculum

Adaptive expertise rarely develops through repetition alone. It develops through structured novelty – intentional variation that requires learners to adapt existing knowledge to new conditions.

Novelty in curriculum design may include:

- Variation in problem conditions
- Unfamiliar application contexts
- Shifts in constraints
- Exposure to counterexamples
- Engagement with competing interpretations

For example:

Routine Task: Solve five equations using the same formula.

Adaptive Task: Determine which mathematical model best explains an unfamiliar real-world scenario and justify your choice under constraints.

Routine Task: Analyze a familiar historical event using known causes.

Adaptive Task: Evaluate how new evidence might alter historical interpretation and defend a revised claim.

Novelty functions best when it is purposeful rather than random. Designed variation strengthens flexible schema and supports transfer across contexts. Without structured novelty, knowledge often becomes brittle.

Teaching Students to Transfer Under Ambiguity

Transfer is the hallmark of adaptive expertise. Yet transfer rarely develops when instruction is highly scripted, narrowly contextualized, performance-driven, or focused primarily on producing answers.

Curriculum that cultivates transfer typically includes:

- Ambiguity: Students must encounter tasks where the path is not predetermined.
- Decision Points: Students select strategies rather than simply follow them.
- Defense: Students articulate why a chosen approach fits the situation.
- Reflection: Students compare alternative strategies and evaluate their effectiveness.

Transfer does not emerge automatically. It develops through repeated opportunities to apply knowledge under changing conditions.

In AI-rich environments, transfer also represents one of the most automation-resistant dimensions of learning. Artificial intelligence can replicate procedures with remarkable speed. Contextual nuance, live adaptation, and interdisciplinary synthesis remain far more difficult to automate.

District assessment systems therefore benefit from greater emphasis on:

- Application under changing conditions
- Defense of reasoning under counterargument
- Revision of thinking based on feedback
- Interdisciplinary reasoning across domains

Without ambiguity, adaptive expertise rarely develops.

Professional Judgment and Instructional Leadership

This shift applies not only to students, but also to teachers and leaders. Compliance-driven systems tend to produce routine expertise – following pacing guides, delivering scripted lessons, implementing programs, and replicating district mandates.

Adaptive systems, by contrast, cultivate professional judgment. Educators exercise judgment when they:

- Adjust instruction in response to student cues
- Diagnose sources of misunderstanding
- Modify tasks strategically
- Design variation intentionally

Artificial intelligence now pressures teacher expertise as well. Scripted curricula, automated planning tools, and generative lesson design systems can create the impression that instruction is sufficient once materials are produced.

Yet adaptive expertise depends on capabilities such as:

- Interpreting instructional context
- Maintaining relational awareness with students
- Exercising ethical discernment in decision making
- Calibrating instruction in real time

Districts that focus exclusively on procedural fidelity often weaken professional agency. Districts that cultivate adaptive expertise strengthen it.

District Tools

System transformation depends on structures that help educators cultivate adaptive expertise. The following coaching tools support curriculum design, assessment analysis, and instructional observation.

Novelty Infusion Planner

Purpose: Introduce structured variation into existing units to strengthen adaptive expertise.

For each unit, identify:

- Core Concept. What must remain across variations?
- Designed Variation: Where can context, constraint, or perspective shift?

Possible variation strategies include:

- Changing the audience
- Altering variables
- Introducing conflicting data
- Adding real-world constraint
- Requiring cross-disciplinary application

Planning questions:

- Does this unit include at least one unfamiliar application?
- Do students make strategic choices?
- Do students defend their reasoning?
- Do students revise their thinking?

When units rely solely on familiar formats, introducing structured novelty can expand adaptive reasoning.

Transfer Disruption Analysis

Purpose: Identify where student transfer breaks down after assessment.

After reviewing assessment results, examine questions such as:

- Can students solve the problem when it is phrased differently?
- Can students explain their strategy without prompts?
- Do students adapt when variables change?
- Can students apply ideas across disciplines?

Common Disruption Patterns:

- Success occurs only when formats remain familiar
- Students rely on surface features rather than underlying concepts
- Students struggle to articulate reasoning
- Minor variation leads to hesitation or confusion

Each disruption often signals limited rehearsal of adaptive thinking. Instructional redesign can then focus on strengthening flexibility.

Adaptive Expertise Observation Rubric

Purpose: Expand instructional observation beyond procedural compliance to include indicators of adaptive thinking.

Observation Domains:

Cognitive Flexibility

- Are students making strategic choices?
- Are students comparing multiple approaches?

Ambiguity Exposure

- Does the lesson include open or non-scripted problem spaces?

Defense & Justification

- Are students explaining their reasoning?

Variation Design

- Are examples varied purposefully?

Revision Culture

- Are alternative strategies discussed and refined?

Each domain can be rated: Emerging. Developing. Established. Embedded.

Observation systems that include these dimensions begin to measure adaptive instructional architecture rather than pacing fidelity alone.

Leadership Reflection

Routine expertise feels safe. It is measurable. It is replicable. It is efficient. Yet efficiency in familiar procedures is no longer a competitive advantage. Artificial intelligence already demonstrates procedural fluency. Human learning therefore turns increasingly toward adaptive expertise.

District leaders can examine the questions this shift raises: Are systems preparing students to execute familiar tasks – or to reason when conditions change? Are teachers being trained to follow scripts – or to exercise professional judgment?

Adaptive expertise is not a luxury. It is the most durable response to intelligent automation. The future will not belong to the fastest procedural performers.

It will belong to those who can:

- Interpret ambiguity
- Adapt strategy
- Transfer insight
- Defend reasoning
- Revise under feedback

This is the expertise worth designing for. And it represents the system-level shift explored throughout this chapter.

Cultivating adaptive expertise among students also depends on adaptive expertise among educators. The next chapter therefore examines how instructional coaching evolves – from feedback on teaching moves toward feedback on learning architecture.

Superintendent Briefing: Preparing Students for Adaptive Expertise

Key Leadership Insight: Artificial intelligence systems now outperform humans at procedural repetition. The enduring advantage of education therefore lies not in speed or accuracy, but in adaptive reasoning under unfamiliar conditions.

District Leadership Implications

1. **Audit curriculum for procedural overemphasis.** Examine whether students primarily practice repetition without variation. When instruction centers on procedural rehearsal alone, routine expertise develops more readily than adaptive expertise.

2. **Embed novelty into assessment design.** Assessments can include unfamiliar conditions where procedures alone are insufficient and students must adapt their reasoning.

3. **Emphasize explanation and defense.** Learning tasks can invite students to justify strategies, compare approaches, and revise reasoning in response to critique.

4. **Develop teacher capacity for intentional variation**. Professional learning can support educators in designing units that incorporate ambiguity, decision points, and contextual shifts.

5. **Align observation systems with adaptive thinking.** Walkthrough tools can examine whether students are adapting strategies rather than simply completing procedures.

Key Question for Leadership Teams: Are our classrooms preparing students to perform familiar tasks – or to think when conditions change?

System Leader Playbook: Cultivating Adaptive Expertise

Artificial intelligence excels at reproducing established procedures. The long-term advantage of human learners therefore lies in their capacity to interpret unfamiliar situations, adapt knowledge, and generate new solutions. District learning systems can be designed to cultivate this form of expertise.

Leadership Moves for Adaptive Learning Systems

1. **Design curriculum that includes structured novelty**. Students encounter problems that require adapting prior knowledge to unfamiliar contexts rather than repeating identical procedures.

2. **Shift assessment toward reasoning under variation.** Assessment tasks can examine how students respond when conditions change, not simply whether they can execute practiced methods.

3. **Encourage explanation and revision.** Learning environments can invite students to justify strategies and refine their reasoning when confronted with new information.

4. **Align observation systems with adaptive thinking**. Classroom observations can focus on student decision-making, problem solving, and strategy adaptation rather than routine completion.

5. **Support professional learning around adaptive instruction.** Educators benefit from opportunities to examine how variation, uncertainty, and cognitive challenge contribute to deeper expertise.

Leadership Question**:** Are our classrooms preparing students to perform familiar tasks – or to adapt when conditions change?

District systems that cultivate adaptive expertise prepare students not only to apply knowledge, but to extend and revise it in new situations.

Chapter 5: Coaching for Design Thinking

Moving from Instructional Moves to Instructional Architecture

System Layer 2: Professional Expertise

Chapter 4 introduced the concept of adaptive expertise – the capacity to revise knowledge and strategy when encountering unfamiliar situations. Adaptive expertise rarely develops in isolation. It grows within professional cultures that examine learning architecture deliberately.

Instructional coaching plays a central role in this process. In many districts, coaching still focuses on visible teaching techniques: questioning strategies, pacing adjustments, or classroom management routines. These practices matter. Yet when coaching remains centered on isolated techniques, improvement tends to remain episodic rather than systemic.

This chapter explores how coaching evolves in AI-rich systems – from feedback on instructional moves to feedback on learning architecture.

Feedback occupies a central role in learning architecture. Hattie's synthesis of educational research found that "the most powerful single modification that enhances achievement is feedback" (2009, p. 173). The impact of feedback, however, depends on the level of instruction it addresses.

Instructional coaching has traditionally emphasized improving teacher "moves": increasing wait time, asking higher-order questions, circulating more intentionally, or clarifying directions. These practices matter. Yet teaching is not simply a collection of techniques – it is the design of a cognitive experience.

In AI-rich systems, coaching increasingly centers on architecture rather than isolated moves.

This chapter therefore reframes instructional coaching as design coaching – anchored in cognitive tension, revision culture, risk survivability, and exposure calibration.

From Moves to Architecture

A move is a moment.

Architecture is a system.

A move might be: "Ask students to justify their answer."

Architecture asks: "Is justification structurally embedded throughout the lesson?"

A move might be: "Encourage participation."

Architecture asks: "Is participation psychologically safe, calibrated, and equitably distributed?"

When coaching focuses only on visible moves, improvement tends to remain episodic. When coaching examines instructional architecture, improvement becomes systemic.

In AI-rich classrooms, architecture ultimately determines whether technology amplifies thinking—or replaces it.

Observing for Design Elements

Coaching for design thinking requires clarity about what to observe. Instructional observation can focus on four core architecture domains.

Cognitive Tension

Cognitive tension functions as the engine of learning. When it is absent, instruction often becomes passive.

Questions for observation:

- Where does the lesson introduce uncertainty?
- Is there a question or problem that requires reasoning?
- Do students generate ideas before receiving explanation?
- Is ambiguity structured and purposeful?

Warning signs:

- Teacher explanation precedes all struggle
- Students complete tasks without verbalizing reasoning
- AI-generated responses are accepted without interrogation

Coaching Prompt: "Where in this lesson did students experience productive uncertainty?"

Revision Culture

Revision signals that learning is iterative. In AI-rich environments, revision becomes even more central to understanding.

Observation questions:

- Are first drafts treated as starting points or final products?
- Is feedback actionable and timely?
- Do students revise publicly as well as privately?
- Is revision normalized rather than punitive?

Warning signs:

- Grading ends the learning cycle
- Feedback focuses on evaluation rather than development
- Students rarely revisit their work

Coaching Prompt: "What opportunities did students have to rethink or refine their thinking?"

Risk Survivability

Intellectual risk is inherent in learning. The critical question is whether that risk is survivable.

Observation questions:

- How are wrong answers handled?
- Are mistakes explored without embarrassment?
- Does participation structure protect student dignity?
- Are quieter voices invited into discussion safely?

Risk survivability often determines whether students stretch – or retreat.

Warning signs:

- Immediate correction shuts down exploration
- Only confident students participate
- Speed is praised over reasoning

Coaching Prompt: "If I were uncertain, would this classroom feel safe to try?"

Exposure Calibration

Exposure refers to how publicly students must think.

Too little exposure can lead to disengagement.
Too much exposure can trigger shutdown.

Effective instruction calibrates exposure carefully.

Observation questions:

- Is thinking scaffolded before public sharing?
- Do students have rehearsal time?
- Does participation progress gradually (partner → small group → whole group)?
- Are high-stakes moments prepared for?

Warning signs:

- Cold-calling without preparation
- Public critique without rehearsal
- Immediate defense without thinking time

Coaching Prompt: "How does this lesson sequence protect dignity while promoting rigor?"

District Tools

Instructional improvement becomes more sustainable when coaching systems focus on lesson architecture rather than isolated teaching techniques. When conversations center on cognitive tension, reasoning opportunities, revision cycles, and transfer, professional learning shifts from evaluating performance to refining design.

The tools below support instructional leaders and coaching teams in examining how lessons are structured to cultivate thinking. By using shared protocols for coaching conversations, lesson analysis, and instructional debriefs, districts can strengthen professional learning cycles that focus on reasoning, revision, and adaptive expertise. These tools help make instructional architecture visible so that coaching consistently reinforces the learning design principles described throughout this book.

Coaching Conversation Protocol

Shifting coaching from instructional moves to instructional architecture benefits from a structured conversation process that centers design rather than isolated techniques.

Step 1: Describe the Architecture. Begin by examining the design of the lesson. Questions to explore:

- "What central cognitive tension did you design?"
- "Where did you anticipate productive struggle?"
- "How was revision incorporated into the lesson?"
- "How was exposure calibrated for students?"

These questions shift reflection from behavior management toward instructional design.

Step 2: Analyze Evidence

Review observable indicators of learning architecture:

- Student talk ratio
- Moments of hesitation or productive struggle
- Evidence of revision
- Attempts at transfer or strategy adaptation
- Emotional climate during discussion

The goal is to examine patterns rather than isolated moments.

Step 3: Identify Structural Levers

Instead of suggesting a new teaching move, explore possible design adjustments.

- "What structural adjustment could increase cognitive tension?"
- "Where might revision be introduced earlier in the sequence?"
- "What exposure scaffold could broaden participation?"

Small architectural shifts often produce more sustainable improvement than isolated techniques.

Step 4: Commit to One Design Shift

Select a single structural adjustment to test in the next lesson cycle.

Possible shifts include:

- Introducing think-before-share routines
- Require explanation before accepting answers
- Adding a revision checkpoint
- Sequencing participation differently to scaffold exposure

Effective coaching commitments remain observable and testable.

Lesson X-Ray Template

The Lesson X-Ray moves beyond surface observation by examining instructional design across five architectural layers: Purpose, Tension, Engagement, Revision, and Transfer. This framework helps coaches analyze how lessons support reasoning rather than simply documenting visible activity.

1. **Purpose Layer**
    - What human capability does the lesson aim to develop?
    - Does the outcome emphasize recall or judgment?
2. **Tension Layer**
    - Where does uncertainty or productive struggle appear?
    - Is struggle structured, time-bound, and supported?

3. **Engagement Layer**
   - Which students are cognitively engaged?
   - Who remains silent?
   - Does silence appear strategic or protective?
4. **Revision Layer**
   - Does the lesson incorporate opportunities for iteration?
   - Is feedback actionable and timely?
   - Is revision visible in student work?
5. **Transfer Layer**
   - Does the lesson require application beyond the immediate example?
   - Are students adapting reasoning to new conditions?

When one of these layers is absent, the instructional architecture may benefit from redesign.

Instructional Design Debrief Tool

Following a coaching cycle, a structured debrief can help teams examine instructional architecture rather than individual performance.

Debrief questions:

1. What worked structurally?

- Where did cognitive tension sustain student thinking?
- Where did students engage in deep reasoning?

2. Where did design weaken?

- Did explanation replace productive struggle?
- Did exposure escalate faster than students could sustain?

3. What design adjustment will be tested next?

- Identify one structural change to examine in the next instructional cycle.

4. What evidence will signal improvement?

Clarify what observable indicators – student talk, reasoning, revision, or transfer – would suggest that the design adjustment is strengthening learning.

The Role of the Instructional Coach in an AI Era

Artificial intelligence tools can now generate lesson plans, question sets, exit tickets, and assessment drafts almost instantly. What they cannot do is interpret the human dynamics that shape learning.

AI systems cannot easily:

- Diagnose relational climate
- Interpret hesitation in student thinking
- Calibrate exposure to intellectual risk
- Design conditions where risk remains survivable
- Notice moments of cognitive shutdown

In this environment, instructional coaches become more – not less – essential. .

The role of the coach increasingly shifts from improving isolated techniques to strengthening instructional design coherence.

Systems that coach for architecture tend to cultivate:

- Cognitive stamina
- Adaptive expertise
- Revision culture
- Ethical reasoning

Systems that focus coaching only on efficiency risk accelerating shallow learning.

Leadership Reflection

Instructional coaching cannot remain compliance-oriented in an AI-rich environment. Increasingly, it becomes design oriented.

Coaching conversations can center on questions such as:

- What kind of thinker did this lesson cultivate?
- Was productive struggle structured?
- Was revision normalized?
- Was intellectual risk survivable?
- Was exposure thoughtfully calibrated?

Design-centered coaching helps ensure that innovation does not outpace cognition.

When coaching strengthens instructional architecture, classrooms become more resilient. In AI-rich systems, resilience – rather than speed – often becomes the defining characteristic of strong learning environments.

The future of coaching lies not in accumulating more tips, but in strengthening instructional design.

Coaching can initiate instructional redesign, but lasting change depends on broader professional structures. The next chapter therefore examines how districts build teacher design capacity across schools so that instructional architecture improves collectively rather than individually.

Superintendent Briefing: Redesigning Instructional Coaching

Instructional coaching systems often emphasize helping teachers refine specific techniques. In AI-rich learning environments, however, coaching increasingly centers on something deeper: the architecture of learning experiences.

District leaders can examine whether coaching systems attend to questions such as:

- Where cognitive tension is introduced
- How student reasoning becomes visible
- Whether revision is structurally embedded
- Whether participation remains psychologically safe
- Whether tasks require transfer under variation

When coaching focuses primarily on teacher behavior, improvement often remains episodic. When coaching examines learning architecture, improvement becomes systemic.

The central leadership question shifts accordingly. Rather than asking only, "Are teachers using effective strategies?" leaders may also ask, "Are learning environments structurally designed to cultivate reasoning?"

District coaching systems therefore benefit from aligning observation and feedback with architectural analysis rather than isolated teaching techniques.

System Leader Playbook: Redesigning Coaching Systems

District redesign benefits from coaching systems that develop instructional design expertise across schools.

1. **Shift coaching observations toward architecture**. Observation protocols can examine cognitive tension, revision structures, and opportunities for transfer.
2. **Transform PLCs into design studios.** Collaborative sessions can focus on analyzing lesson architecture rather than reviewing pacing or materials alone.
3. **Train coaches as design analysts.** Coaches deepen their capacity to identify intrinsic cognitive load, productive struggle, and moments where reasoning becomes visible.
4. **Align coaching feedback with district learning goals.** Feedback can reinforce the same intellectual priorities reflected in curriculum and assessment.
5. **Institutionalize redesign cycles.** Teachers test one architectural shift at a time and examine its effects on student reasoning.

When coaching systems develop design expertise, professional learning becomes continuous rather than episodic.

Chapter 6: Building Teacher Design Capacity

From Individual Excellence to System Architecture

System Layer 2: Professional Expertise

Instructional architecture alone does not transform learning systems. Even thoughtfully designed curriculum and well-structured coaching depend on educators who can interpret student thinking and redesign learning experiences continuously.

Chapters 4 and 5 examined key foundations of professional expertise—adaptive judgment and coaching for instructional architecture. This chapter extends that work by exploring how districts develop teacher design capacity across entire systems.

When design expertise remains isolated within individual classrooms, improvement often remains fragile. When it becomes part of a shared professional culture, instructional redesign becomes sustainable.

Professional expertise therefore depends not only on individual skill but also on structures that make design thinking visible, collaborative, and iterative.

Beliefs about intelligence shape learning behavior. As Dweck explains, "students' beliefs about their intelligence have a powerful impact on their learning and achievement" (2006, p. 6). When schools communicate that ability is fixed, students may protect identity by avoiding challenge. When schools communicate that ability develops through effort and strategy, students tend to persist longer in the face of difficulty.

Many strong schools are sustained by exceptional individuals – the master teacher, the innovative early adopter, the charismatic instructional leader. Yet systems built on heroics remain fragile. When excellence depends primarily on individual instinct rather than structural support, coherence often dissolves as personnel shifts.

In an AI-saturated era – where curriculum tools, lesson generators, and automated planning systems proliferate – the central challenge is rarely a shortage of ideas. More often, systems lack design coherence.

This chapter argues for a decisive shift: System capacity over individual heroics.

Design capacity becomes most durable when it is institutionalized across the system.

The Problem with Heroic Teaching

Heroic teaching is often inspirational, energetic, and innovative. It is also frequently unsustainable.

Heroic teachers often:

- Design extensively in isolation
- Carry cognitive load without collaboration
- Experiment without systemic support
- Burn out quietly

Meanwhile, surrounding classrooms may operate under entirely different instructional philosophies.

The result can be inconsistent student experiences, uneven cognitive demand, and fragmented identity formation.

System-level leadership can therefore examine a critical question: *If our strongest teachers left tomorrow, would our instructional design philosophy endure?*

If the answer is no, the system remains personality dependent. Durable capacity emerges when expertise is embedded in structures rather than individuals.

Professional Learning Redesign

Traditional professional learning often centers on tool adoption, compliance updates, program rollouts, and strategy demonstrations. These approaches tend to build procedural fluency but rarely develop design judgment.

In an AI-rich environment, professional learning increasingly shifts from asking "How do we implement this?" to asking "How do we design learning architecture intentionally?"

Effective professional learning experiences often include:

- Analysis of lesson architecture
- Examination of cognitive load
- Practice with structured novelty
- Rehearsal of exposure calibration
- Refinement of feedback architecture

Developing design expertise requires opportunities for educators to practice instructional design—not simply discuss it.

Professional learning environments therefore function more like studios: iterative, reflective, and collaborative rather than presentation driven.

PLC as Design Lab (Not Data Meeting)

Many Professional Learning Communities (PLCs) gradually shift toward score reviews, pacing checks, administrative updates, and surface-level sharing.

Data remains important. Without design analysis, however, data review often becomes reactive rather than developmental.

A PLC functioning as a design lab explores questions such as:

- What cognitive tension was intentionally designed?
- Where did transfer break down?
- How did revision operate within the unit?
- What kind of thinker did this learning experience cultivate?

Instead of asking, "Why did scores drop?"

Teams can ask, "Where did the instructional architecture fail to support reasoning?"

Design labs encourage practices such as:

- Prototyping new lesson structures
- Testing variations in instructional design
- Revising based on evidence from classrooms
- Building coherence across classrooms

In this model, the PLC becomes a central engine of adaptive expertise.

Iterative Instructional Prototyping

Instructional design improves through iteration rather than perfection. When teachers feel pressure to "get it right" the first time, experimentation often diminishes and innovation slows.

Design thinking treats instruction as an evolving process:

Prototype → Test → Reflect → Redesign

Educators benefit from opportunities to:

- Pilot design shifts
- Analyze student responses
- Adjust exposure levels
- Refine points of cognitive tension

Prototyping reframes "unsuccessful lessons" as sources of information. It encourages experimentation within a structured design cycle.

Without iteration, instructional design tends to stagnate. Without psychological safety, experimentation becomes unlikely.

Supporting design thinking therefore involves normalizing instructional testing cycles across the system.

Feedback Loops

Iteration becomes meaningful when evidence moves quickly back to those designing instruction. Feedback systems therefore help educators see how instructional architecture is functioning in practice.

Strong design capacity relies on feedback loops operating at multiple levels.

Classroom Level

- Student exit data
- Observation notes
- Structured reflection protocols

PLC Level

- Cross-classroom comparisons
- Analysis of cognitive tension calibration
- Mapping of transfer disruptions

District Level

- Instructional coherence audits
- Alignment reviews with design principles
- Vertical articulation across grade levels

Effective feedback centers on instructional architecture rather than individual personality.

The guiding question shifts accordingly. Instead of asking, "Did the teacher perform well?"

Leaders ask, "Did the instructional design function as intended?"

System change depends on infrastructure. Instructional redesign rarely sustains itself through individual enthusiasm alone; it develops when repeatable structures make experimentation, reflection, and refinement routine. The models below illustrate several ways districts can institutionalize design capacity.

The 90-Day Design Cycle

Purpose: Translate improvement goals into iterative instructional redesign.

Improvement efforts often stall when reflection remains disconnected from classroom implementation. The 90-Day Design Cycle creates a predictable rhythm that moves teams from diagnosis to redesign and back to analysis.

Each cycle includes four phases.

Phase 1: Diagnose (Weeks 1–3)

- Identify architectural weakness
- Conduct coherence audit
- Map cognitive load patterns

Phase 2: Design (Weeks 4–6)

- Develop revised unit prototypes
- Introduce structured novelty
- Embed revision checkpoints
- Calibrate exposure levels

Phase 3: Implement (Weeks 7–10)

- Teach the redesigned unit
- Collect observational evidence
- Capture artifacts of student reasoning

Phase 4: Reflect & Refine (Weeks 11–12)

- Analyze evidence of transfer
- Identify points where cognitive tension collapsed
- Refine the instructional architecture

The 90-day cycle helps prevent improvement drift by anchoring change in design iteration.

Redesign Studio Model

Purpose: Transform PLCs into instructional design laboratories.

Professional Learning Communities often concentrate on logistics or data review rather than instructional architecture. The Redesign Studio reframes collaborative time as a structured design lab.

Studio Structure

A. Present the Architecture

A teacher introduces the unit design goals and planned sources of cognitive tension.

B. X-Ray the Lesson

Colleagues analyze the architecture of the lesson, including:

- Intrinsic cognitive load
- Novelty infusion
- Risk survivability
- Evidence of transfer

C. Prototype Adjustment

The team proposes structural redesigns rather than isolated teaching moves.

D. Implementation Commitment

The teacher tests one architectural shift in the next instructional cycle.

Studios cultivate a shared design language, reduce professional isolation, and normalize iteration.

Cognitive Apprenticeship Model for Teachers

Purpose: Develop instructional expertise through structured modeling, coaching, and reflective practice.

Instructional expertise rarely develops through workshops alone. Teachers, like students, learn complex practices through guided participation with more experienced practitioners.

Cognitive apprenticeship typically includes:

- Modeling
- Coaching
- Scaffolding
- Articulation
- Reflection
- Fading

District Application

Modeling: Lead teachers demonstrate design thinking during planning.

Coaching: Instructional coaches provide architecture-focused feedback.

Scaffolding: New teachers co-design units before leading independently.

Articulation: Teachers explain the reasoning behind design decisions.

Reflection: Post-lesson debriefs examine how the architecture functioned.

Fading: As design fluency increases, scaffolds gradually decrease, making professional growth visible and sustainable.

Culture Shift: From Compliance to Craft

Design capacity tends to flourish in professional cultures that value experimentation over perfection, reflection over blame, and collaboration over isolation. Compliance-driven systems often suppress innovation because deviation appears risky. Design-oriented systems treat iteration as a form of professional craft.

Leadership conversations can explore questions such as:

- Do existing structures incentivize safe replication – or thoughtful redesign?
- Do teachers feel psychologically safe to prototype new approaches?
- Is instructional risk treated as a learning opportunity or a liability?

Design capacity rarely grows through workshops alone. It develops through culture.

Leadership Reflection

Artificial intelligence does not eliminate the need for teachers. It challenges the assumption that procedural replication alone is sufficient.

In this environment, teaching increasingly centers on design. Coaching evolves toward analysis of learning architecture. PLCs function more like design studios. Districts operate as systems of apprenticeship.

The leadership question shifts accordingly.

It is no longer simply: "Do we have strong teachers?"

It becomes: "Have we built a strong design system?"

When design capacity is institutionalized, innovation spreads, coherence strengthens, burnout decreases, and professional judgment deepens. System capacity begins to outlast individual heroics.

In an era defined by acceleration, sustainability becomes a critical design advantage. The future of teaching depends less on extraordinary individuals and more on systems that make design expertise an ordinary professional expectation.

Superintendent Briefing: Institutionalizing Design Expertise

Professional expertise rarely sustains improvement when it depends solely on exceptional individuals. More durable progress emerges when systems create conditions that allow educators to design, test, and refine learning architecture collectively.

Effective district structures often support:

- Collaborative lesson design
- Classroom observation focused on learning architecture
- Shared language around reasoning and transfer
- Iterative redesign cycles
- Psychological safety for experimentation

When expertise is treated primarily as individual talent, innovation tends to remain isolated. When expertise becomes a shared system expectation, instructional quality strengthens across schools.

The leadership question shifts accordingly: Are we relying on strong teachers – or building strong design systems?

Districts that institutionalize design expertise often see reduced burnout, stronger instructional coherence, and deeper professional judgment across classrooms.

System Leader Playbook: Building Professional Design Capacity

District redesign depends on more than strong curriculum. It also relies on the professional judgment of educators who intentionally design learning environments. System leaders therefore benefit from establishing structures that develop adaptive expertise across the organization. Leadership moves can include:

1. **Redesign professional learning.** Shift professional learning from strategy workshops toward design studios where teachers analyze lesson architecture, cognitive tension, and opportunities for transfer.
2. **Align coaching with instructional architecture**. Instructional coaching can examine elements such as cognitive tension, revision culture, risk survivability, exposure calibration, and transfer opportunities. Observation protocols emphasize thinking structures rather than pacing compliance.
3. **Protect collaborative design time.** Adaptive expertise develops through iteration. Systems that cultivate it often protect regular time for unit redesign, transfer analysis, assessment conversion, and lesson X-rays.
4. **Develop design language across the district.** Establishing common vocabulary helps teams discuss learning architecture with clarity. Examples include cognitive tension, productive struggle, adaptive expertise, evidence of transfer, and revision culture. Shared language strengthens instructional coherence.
5. **Institutionalize iteration cycles**. Many districts support instructional redesign through structured improvement cycles such as: Diagnose → Redesign → Implement → Reflect → Refine. When redesign cycles become routine, professional growth begins to function as system architecture rather than isolated innovation.

PART III – ETHICAL AI INTEGRATION

Chapter 7: The Moral Compass of AI in Schools

Designing With Integrity, Not Just Innovation

AI integration in schools is not primarily a technical decision. It is a moral one.

Ethical discussions surrounding artificial intelligence increasingly focus on responsibility and human oversight. Scholars emphasize that AI governance should ensure that "humans remain meaningfully responsible for the outcomes of automated systems" (Floridi et al., 2018, p. 695). Educational technologies are never neutral instruments; they "reflect and reinforce particular social and political priorities" (Selwyn, 2019, p. 15).

Across the country, school systems are introducing a growing range of AI-enabled tools, including:

- AI writing assistants
- Automated grading systems
- Predictive analytics dashboards
- Behavior monitoring software
- Plagiarism detection systems
- Adaptive learning platforms

The language surrounding adoption often emphasizes efficiency, innovation, personalization, and acceleration. Yet efficiency alone does not constitute an educational value. School systems exist not merely to accelerate production but to cultivate intellectual and human formation.

For system leaders, the integration of AI therefore becomes a question of institutional architecture. Decisions about tools, policies, and data systems shape how judgment is exercised, how responsibility is distributed, and how human agency is preserved within technological environments.

This chapter extends the ethical foundations introduced in *AI and the Art of Teaching* and reframes AI integration through a system design lens.

The central question is not simply: Can we use AI?

The more consequential question for educational systems becomes: Under what conditions does its use strengthen – rather than weaken – the human purposes of schooling?

Surveillance vs. Support

One of the earliest ethical tensions in AI integration concerns visibility.

Contemporary AI systems can track student keystrokes, monitor browsing behavior, analyze engagement time, flag "suspicious" writing patterns, and generate predictions about academic risk.

These tools are often introduced as mechanisms for support. In practice, many function as systems of surveillance.

Surveillance architectures reshape school culture. When monitoring becomes the dominant mechanism for understanding student behavior, trust weakens. Anxiety increases. Suspicion becomes normalized. Responsibility gradually shifts away from professional relationship and toward technological oversight.

Support-oriented systems operate differently. They use information to strengthen learning rather than to control behavior. Within support-centered environments, data systems are designed to:

- Increase transparency
- Prioritize growth over detection
- Encourage student ownership
- Protect student dignity

When data becomes the primary lens through which students are interpreted, the relational foundation of teaching begins to weaken. Information should inform professional judgment, not replace it.

For system leaders, the ethical question is therefore architectural rather than technical. AI tools shape how visibility operates inside schools: what is monitored, who interprets information, and how authority is exercised.

Leaders evaluating new technologies may consider questions such as:

- Does this tool strengthen relational trust, or substitute monitoring for relationship?
- Does the system assume student integrity, or organize around suspicion?
- Are structures designed to support accountability, or to amplify fear?

Ethical AI integration ultimately depends on preserving student dignity within technologically complex environments. The purpose of educational technology is not to replace human relationship, but to strengthen the conditions under which it can flourish.

Efficiency vs. Formation

AI's greatest strength is speed. It drafts instantly, sorts rapidly, analyzes at scale, and responds without delay. Human formation, however, unfolds more slowly.

Formation develops through struggle, revision, dialogue, reflection, and sustained engagement with uncertainty. These processes may appear inefficient when measured through the lens of productivity, yet they are precisely the conditions under which judgment matures.

When educational systems organize primarily around efficiency, they may unintentionally compress the experiences that cultivate expertise. The question for system leaders is therefore not whether efficiency has value, but how efficiency interacts with the deeper purposes of learning.

Consider common examples:

- Automated grading may accelerate feedback – but does it deepen student reasoning?
- AI-generated lesson plans may reduce planning time – but do they strengthen teachers' design expertise?
- Instant summaries may shorten reading – but do they cultivate analytical thinking?

These questions highlight an architectural tension within modern schooling. Systems designed primarily for speed risk narrowing the intellectual work that sustains deep learning.

For districts committed to intellectual formation, preserving certain forms of productive slowness may be necessary. Time for revision, dialogue, and conceptual exploration often functions as part of the learning infrastructure rather than as a barrier to progress.

Not all friction represents inefficiency. In many cases, friction supports the development of judgment, insight, and professional expertise.

Equity Implications

AI does not enter neutral terrain. It enters educational systems already shaped by historical and structural inequality. Equity considerations therefore extend beyond simple access to technology. They include how AI tools interact with opportunity, authority, and protection within existing institutional structures. For system leaders, several questions can surface when evaluating AI adoption:

- Access: Do all students have meaningful opportunities to engage with these tools, or do AI systems primarily amplify advantages already held by well-resourced learners?
- Skill Amplification: Do AI tools extend the thinking of students who already possess strong academic foundations, while further marginalizing learners who are still developing essential skills?
- Language Bias: Are multilingual students more likely to be flagged by detection systems trained primarily on dominant language patterns?
- Data Representation: Are predictive models built on historical datasets that may reproduce existing disparities in opportunity or discipline?

Without careful examination, AI systems may widen the very gaps schools seek to close.

Equity cannot be secured through access alone. It requires sustained attention to how technological systems distribute benefit and risk across student populations. Within any data-driven environment, certain questions remain essential: Who is empowered? Who is monitored? Who is protected? Who is exposed?

From a system design perspective, ethical AI integration therefore involves more than selecting tools. It involves examining how those tools interact with the architecture of opportunity inside schools. The central equity question becomes difficult but necessary: Are we solving instructional challenges – or embedding existing inequities into automated systems?

Bias Amplification

AI systems are trained on large datasets. These datasets reflect historical and societal patterns – including bias. As a result, AI systems do more than process information; they may reproduce and amplify existing inequities embedded within the data on which they are trained.

In educational contexts, bias may surface across a range of AI-enabled tools, including language generation systems, risk prediction models, automated scoring platforms, and recommendation algorithms.

When districts adopt such systems without examining the assumptions embedded in their design, bias can become integrated directly into institutional decision-making processes.

Potential consequences may include:

- Reinforcing existing stereotypes
- Over-identifying certain student groups for intervention
- Misinterpreting cultural or linguistic communication patterns
- Encoding inequitable assumptions into policy and practice

Bias amplification is rarely obvious. It often operates beneath claims of objectivity and the perceived neutrality of mathematical models.

For system leaders, the challenge therefore becomes architectural rather than purely technical. AI systems influence how information is interpreted, how risk is categorized, and how decisions are distributed across institutional structures.

Responsible governance often involves establishing practices that promote transparency and accountability in technological systems. Districts evaluating AI tools may consider expectations such as:

- Transparency regarding model training data and design assumptions
- Independent evaluation of potential bias in system outputs
- Clear policies governing data use, interpretation, and oversight

Automation alone does not guarantee fairness. Technologies that influence educational decision-making benefit from careful scrutiny, professional judgment, and transparent governance structures.

Transparency Norms

Students and families benefit from clear understanding of how technological systems influence learning environments.

AI tools increasingly shape grading, feedback, scheduling, and intervention decisions. When these systems operate without visibility, trust within the school community can begin to weaken. Transparency therefore functions not merely as a procedural requirement but as an essential element of ethical system design.

Meaningful transparency typically includes several elements:

- Clear explanations of how AI tools are used in learning processes and decision-making structures
- Disclosure of what data is collected, how it is stored, and how it may be used
- Explicit boundaries governing monitoring or surveillance technologies
- Clearly defined human authority for final decisions

Transparency strengthens trust within educational communities. Hidden automation tends to erode it.

When AI systems influence educational experiences, students and families benefit from understanding several key aspects of system operation:

- When AI tools are involved in educational processes
- What role those tools play in shaping feedback or decisions
- Who retains final responsibility for judgment and action

Ethical educational systems do not obscure decision-making behind automated processes. Instead, they acknowledge how technology participates in institutional decisions and maintain clear human accountability for the outcomes that follow.

District Tools

District AI committees and leadership teams benefit from structured frameworks when evaluating new technologies. Before purchasing, piloting, or expanding AI tools used for instruction, grading, monitoring, or decision-making, system leaders may examine how those tools interact with the district's instructional architecture, ethical commitments, and community trust.

The following tools support reflective evaluation prior to adoption and throughout implementation.

The Ethical AI Decision Compass

Purpose: Evaluate whether proposed AI tools support learning, preserve student dignity, and align with district instructional priorities before adoption or expansion.

Before implementing or scaling any AI tool, leadership teams may explore four guiding questions.

1. **Does this tool reduce or replace thinking?**

Some reductions in cognitive effort are productive. Removing extraneous burden can allow learners to focus on conceptual work. Replacement, however, removes the very intellectual effort that builds understanding.

When evaluating tools, leaders may consider:

- What cognitive work remains human?
- Where do students still wrestle with ideas?
- Is first-draft thinking preserved?

Tools that consistently substitute for reasoning may weaken instructional integrity.

2. **Does it widen or narrow identity risk?**

Identity risk refers to the vulnerability students experience when thinking publicly. Certain technologies increase exposure by flagging students, comparing performance publicly, or identifying errors algorithmically. Others provide scaffolding, private rehearsal, or supportive feedback that encourages participation.

Leaders may therefore ask: Does this tool protect dignity, or increase public exposure?

3. **Does it privilege speed over depth?**

Many AI tools promise acceleration. Yet depth of learning often develops through pause, iteration, defense, and reflection.

A useful leadership question becomes: Would the learning outcome improve if the process unfolded more slowly?

If so, acceleration may conflict with instructional goals.

4. **Who benefits? Who is exposed?**

AI systems redistribute efficiency, authority, and risk across an organization.

Leadership teams may map stakeholders: Students. Teachers. Administrators. Families. Vendors.

Questions may include:

- Who gains efficiency?
- Who loses agency?
- Who assumes risk if the system fails?
- Whose data supports the system?

Ethical adoption often depends on ensuring that benefits and risks are distributed thoughtfully rather than concentrated unevenly.

AI Ethics Audit

Purpose: Surface ethical risks within existing or proposed AI systems and evaluate how those systems influence learning, decision-making, and student dignity.

Ethical concerns rarely appear in product descriptions. They typically emerge in how tools reshape daily instructional practice.

An ethics audit may explore questions such as:

- Does the tool replace student thinking or support it?
- Does it increase surveillance or strengthen relational trust?
- Does it prioritize speed over reasoning and intellectual formation?
- Does it introduce new equity risks or amplify existing disparities?

Ethical review functions best as an ongoing process rather than a one-time event. Technologies evolve, and so do their impacts on classrooms and school systems.

Responsible districts therefore treat ethical evaluation as part of their governance architecture.

AI Classroom Impact Mapping

Anticipate how AI tools may influence student cognition, teacher expertise, and equity across the instructional system before implementation.

Leadership teams can map potential risks alongside protective safeguards.

Domain	Risk	Safeguard
Student cognition	AI substitutes for reasoning	Require explanation or defense of ideas
Teacher expertise	AI generates instructional plans	Include reflective design review
Equity	Biased detection	Conduct output audits and equity reviews

Impact mapping helps districts recognize where safeguards may be necessary before classroom integration occurs.

AI Transparency Protocol

Purpose: Clarify how districts communicate with families, students, and staff about the role AI systems play in educational decisions.

Transparency supports trust when communities understand:

- what data is collected
- who accesses that data
- how algorithmic outputs influence decisions
- how families are informed about AI use

When AI tools influence grading, monitoring, or decision-making, visibility into system operation helps preserve accountability within the educational community.

Transparency is therefore less about disclosure alone and more about maintaining clear human responsibility for decisions influenced by technology.

AI Governance Review Committee

Purpose: Provide structured oversight of district AI decisions by incorporating instructional, ethical, technical, and community perspectives.

Membership may include district leaders, classroom teachers, technology staff, students, families, and data privacy specialists.

Responsibilities may include:

- reviewing proposed AI tools
- applying the Ethical AI Decision Compass
- examining potential bias or privacy concerns
- communicating policies and expectations to the community

Governance structures help ensure that AI adoption reflects educational priorities rather than technological momentum.

System Implementation

Ethical AI integration rarely succeeds when approached informally. Because AI tools influence instructional practice, student data, and decision-making systems, thoughtful implementation benefits from clear organizational structures.

Rather than relying on isolated decisions, districts often develop governance processes that guide evaluation, adoption, and oversight across the system.

Implementation frameworks commonly include elements such as:

- AI review or governance committees that examine new tools before adoption
- Transparent evaluation and adoption protocols for instructional technologies
- Bias and equity review processes for AI-enabled systems
- Communication plans that inform families, staff, and students about AI use
- Periodic ethical reviews that revisit the impact of implemented tools

Within this structure, tools such as the Ethical AI Decision Compass function less as abstract guidance and more as part of the district's operational review process.

When ethical reflection becomes embedded within implementation routines, principles begin to function as system architecture rather than philosophical statements.

Policy and practice then move into closer alignment. Ethical commitments become visible not only in written guidelines, but in the decision pathways through which technologies enter classrooms.

Leadership Reflection

Artificial intelligence is neither inherently beneficial nor inherently harmful. In practice, it tends to reflect the priorities embedded within the systems that adopt it.

When school systems prioritize efficiency, AI often accelerates efficiency. When systems emphasize compliance, AI can automate compliance. When systems center intellectual and human formation, AI may be structured to support those purposes – within carefully designed boundaries.

The central tension of AI integration is therefore less technical than ethical. The question facing educational leaders is not simply how to deploy new tools, but what kinds of learning environments those tools ultimately reinforce.

In this sense, AI introduces more than technological capability into schools. It introduces new forms of institutional power – the power to monitor behavior, automate judgment, and accelerate decisions that once depended on human interpretation.

Educational systems therefore face an architectural choice: whether these capacities will be organized to control learning or to strengthen it.

Technology that substitutes for thinking gradually weakens the purposes of education.
Technology that protects human judgment can strengthen them.

For leaders, the defining question is no longer whether artificial intelligence will exist within schools. In many contexts, it already does.

The deeper question is whether our institutional structures ensure that human dignity, professional judgment, and intellectual development remain at the center of educational life.

When systems are designed with those commitments in mind, technology becomes a tool that supports learning rather than directing it.

Superintendent Briefing: Ethical AI Governance

Artificial intelligence tools are entering schools rapidly, often through individual classroom experimentation or vendor partnerships. Without clear ethical frameworks, districts may adopt systems that unintentionally weaken student reasoning, expand surveillance practices, or reproduce historical patterns of bias.

For district leaders, AI governance therefore becomes a matter of institutional design rather than isolated technology decisions. Effective governance structures often articulate guiding principles that shape how tools are evaluated, implemented, and monitored across the system.

Common governance priorities include:

- protecting student dignity within data-driven environments
- maintaining clear human authority in educational decision-making
- examining AI tools for potential bias or inequitable outcomes
- prioritizing learning designs that preserve student reasoning rather than automate it

When these principles guide adoption processes, technology can function as a support for the intellectual work of students and educators rather than a substitute for it.

Ethical AI governance typically includes transparent evaluation protocols, clear communication with families and communities, and ongoing review of how emerging technologies influence classroom practice and student learning.

The leadership question ultimately becomes architectural:

Are we integrating AI in ways that deepen learning – or in ways that simply accelerate automation?

System Leader Playbook: Ethical AI Integration

District leaders shape how artificial intelligence functions within educational systems. Ethical integration depends less on individual tools and more on the governance structures that guide their evaluation, adoption, and oversight.

Several leadership practices help align AI use with instructional priorities and community trust:

1. **Establish an AI Ethics Review Committee.** Create a structured process for examining AI tools before adoption, including consideration of instructional value, student dignity, and potential unintended consequences.

2. **Develop transparent AI use protocols.** Document how AI tools collect data, how algorithms influence feedback or decisions, and where human authority remains in the decision-making process.

3. **Conduct bias and equity reviews.** Examine how AI systems perform across diverse student populations, including multilingual learners and historically marginalized groups.

4. **Protect student reasoning in instructional design.** Ensure classroom tasks continue to require explanation, analysis, and defense of ideas even when AI tools are available.

5. **Communicate clearly with families and communities.** Provide accessible explanations of how AI tools are used in learning environments and how student data is safeguarded.

When governance structures make these practices routine, technology becomes integrated within the instructional system rather than operating outside it.

Ethical AI leadership ultimately ensures that technological innovation strengthens learning environments rather than reshaping them in unintended ways.

Chapter 8: AI Governance & Policy Design

From Panic Response to Principled Architecture

Chapter 7 explored the ethical principles that guide artificial intelligence in schools, and this chapter turns to the structures that help those principles operate in practice. Ethical commitments gain meaning when they are reflected in the policies, review processes, and communication systems that shape daily decision-making across a district.

Artificial intelligence is already present in most school systems. In some cases, it appears through formal district platforms. In others, it emerges more quietly through everyday professional practice. Teachers may use AI tools while planning instruction. Administrators may rely on them when summarizing reports or organizing data. Specialists may explore translation features. The central question for district leadership is therefore less about permission and more about stewardship: how AI use can be guided responsibly within the learning system.

Educational technologies are rarely neutral. As Selwyn notes, digital systems often "reflect and reinforce particular social and political priorities" (2019, p. 15). Governance therefore extends beyond managing misuse or enforcing restrictions. It involves designing conditions that clarify expectations, maintain transparency, support professional judgment, and align technology use with educational purposes.

Within a learning architecture framework, governance functions as a stabilizing layer. It helps districts clarify boundaries for AI use, establish transparent decision processes, examine equity implications, and ensure that technology supports reasoning rather than replacing it.

This chapter offers practical structures districts may use when developing AI governance policies.

When governance structures are thoughtfully designed, AI integration can remain aligned with the intellectual and relational goals of schooling.

Academic Integrity: Evolution, Not Elimination

Traditional academic integrity frameworks developed in an era when misconduct was typically defined by plagiarism, copying, or unauthorized collaboration. Artificial intelligence introduces a more complex landscape. AI-generated responses may be original in wording, structurally coherent, and technically unplagiarized, yet still represent reasoning that the student did not produce.

This shift invites districts to reconsider how integrity is defined. Rather than focusing exclusively on whether work was copied, integrity frameworks increasingly examine whether reasoning is authentically owned.

In this context, academic integrity emphasizes elements of student work:

- ownership of reasoning
- transparency regarding tool use
- the ability to explain and defend thinking
- evidence of revision and intellectual development

The goal of governance is to clarify the conditions under which AI tools can support learning while preserving the intellectual work that remains central to education.

Policies may therefore articulate where AI assistance is appropriate, where AI use is restricted, and which forms of work must remain primarily human authored. When expectations are explicit, integrity shifts away from prohibition toward accountability and transparency.

In AI-rich learning environments, integrity frameworks align closely with the instructional design principles introduced earlier in this book: explanation, transfer, revision, and the defense of reasoning. When these elements are built into learning architecture, student thinking becomes visible in ways that strengthen both learning and integrity.

AI Transparency Expectations

Ethical AI governance depends not only on rules, but on clarity. When expectations about artificial intelligence remain implicit, confusion and suspicion can follow. Transparency helps districts establish shared understanding about how AI tools are used, what boundaries exist, and where human judgment remains central. Within a governance framework, transparency clarifies expectations across multiple roles in the learning system.

Students benefit from understanding:

- when AI tools may support learning
- when AI use is restricted
- when disclosure of AI assistance is expected

Teachers benefit from clarity regarding:

- whether AI tools may assist with grading preparation or feedback drafting
- when AI-supported feedback is appropriate
- what forms of documentation or disclosure are expected

Administrators benefit from shared visibility into:

- which AI tools are approved for district use
- what data privacy agreements are in place
- what oversight structures guide responsible implementation

When expectations are communicated openly, uncertainty tends to decrease and professional judgment becomes easier to exercise. In the absence of transparency, policies can appear reactive or punitive. With clear communication, districts are better positioned to sustain trust while integrating emerging technologies responsibly.

Professional Trust vs. Compliance Policing

AI governance does more than regulate technology. It also shapes the professional culture in which educators and students operate. District policies can signal whether AI integration will be approached primarily through monitoring and enforcement or through professional judgment supported by clear expectations.

In practice, governance systems tend to lean toward one of two orientations.

Compliance-Oriented Model includes surveillance-based detection tools, suspicion-driven enforcement processes, public exposure of violations, and expanded monitoring of student activity.

Professional Trust Model

- clearly articulated guidelines for AI use
- explicit boundaries around tools and tasks
- expectation of transparent disclosure
- emphasis on explaining and defending reasoning

Compliance-oriented systems may reduce visible misuse in the short term. Over time, however, they can weaken trust and discourage open conversation about how tools are actually used. Governance systems grounded in professional trust approach the issue differently. Expectations remain explicit, accountability remains present, and the emphasis shifts toward intellectual ownership rather than simple rule enforcement.

District leadership conversation returns to a central question: What kind of culture are our policies reinforcing – one rooted in suspicion, or one that encourages responsible judgment?

Trust does not remove boundaries. It makes those boundaries visible, understandable, and sustainable within the learning system.

AI in Grading

Grading decisions occupy a particularly sensitive space within AI governance. Assessment not only evaluates academic work but also reflects professional judgment, contextual understanding of student learning, and the relational knowledge teachers develop over time. As AI tools begin to assist with feedback and evaluation tasks, districts benefit from clarifying how these tools interact with educator expertise.

Several considerations may arise when AI is introduced into grading processes. These include potential bias within automated scoring systems, overreliance on machine-generated feedback, loss of relational nuance about student effort or growth, and diminished visibility of teacher judgment in evaluation decisions.

At the same time, AI tools can provide useful support when applied thoughtfully. In many contexts, they may assist educators by identifying patterns in student responses, suggesting possible feedback language, and highlighting missing components in student work. Used in this way, AI functions as an analytical aid rather than a decision-maker.

Governance frameworks therefore tend to clarify several expectations around grading practices:

- AI tools may support the preparation of feedback or analysis of student work
- teachers retain final authority over grading decisions
- AI-generated feedback is reviewed by the educator before being shared with students

These boundaries help ensure that efficiency does not displace professional discernment. When grading remains anchored in educator judgment, AI tools can contribute to the process without replacing the relational and contextual expertise that effective evaluation requires.

AI in IEP Documentation

Special education documentation introduces additional considerations for AI governance. Individualized Education Plans (IEPs) require careful attention to student needs, family voice, and legal protections. As AI tools begin to assist with documentation tasks, districts benefit from clarifying how these tools may support efficiency while preserving individualized decision-making and student privacy.

In some contexts, AI tools may assist by drafting preliminary goal language, summarizing meeting notes, organizing documentation structures, and suggesting phrasing for progress reports. Used thoughtfully, these functions can reduce administrative burden and free time for direct student support.

At the same time, several risks accompany AI-assisted documentation:

- goals becoming overly generalized rather than individualized
- loss of nuance about a student's specific learning profile
- potential exposure of confidential student data
- misrepresentation of student or family voice in documentation

Governance frameworks therefore tend to clarify several protective expectations:

- AI-generated documentation receives human review and verification before use
- identifiable student information is not uploaded into unsecured platforms
- data privacy practices remain consistent with district and legal requirements

Within these guardrails, AI may assist with drafting and organization while individualized educational judgment remains firmly in human hands.

AI and Multilingual Learners

Artificial intelligence tools can play a meaningful role in supporting multilingual learners when integrated thoughtfully into instructional systems. Language support technologies have the potential to expand access to academic content, reduce linguistic barriers, and help students participate more fully in classroom learning.

In many learning environments, AI tools may assist multilingual learners by providing instant translation of academic language, vocabulary scaffolding and definition support, sentence frame generation for academic writing, reading level adjustments that improve accessibility. These forms of support can help students engage with complex ideas while continuing to develop language proficiency. At the same time, several equity considerations accompany AI-supported language tools:

- language simplification that may unintentionally reduce conceptual depth
- cultural or linguistic bias embedded in generated phrasing
- increased monitoring or surveillance of language patterns
- multilingual writing being incorrectly flagged as "AI-generated"

Governance frameworks therefore benefit from clarifying expectations that protect linguistic diversity within AI-supported learning environments. Policies may affirm that AI tools can function as legitimate language scaffolds, language support technologies are not automatically treated as integrity violations, and multilingual expression differences are not misinterpreted as misconduct.

When governance structures recognize these distinctions, AI integration can support language development while maintaining fairness and dignity for multilingual learners. Ethical AI governance ultimately aims to protect – rather than penalize – linguistic diversity within the learning system.

District Tools

System-level governance becomes sustainable when districts rely on clear tools rather than informal interpretation. The structures below provide practical mechanisms for clarifying expectations, evaluating technology responsibly, and maintaining transparency across the learning system.

These tools help leadership teams move from philosophical agreement about ethical AI use toward consistent governance practices that support reasoning, equity, and professional judgment.

AI Integration Policy Blueprint

Purpose: Establish a clear district framework for how artificial intelligence tools may be used in learning environments while preserving student reasoning and professional judgment.

A policy blueprint helps districts communicate the philosophical foundations and operational boundaries of AI integration. When expectations are articulated clearly, educators and students can use emerging technologies with greater confidence and consistency.

1. Purpose Statement

District policy typically begins with a clear statement of intent. This statement communicates that:

- AI tools may support learning processes
- AI tools do not replace student reasoning
- professional judgment remains central to instructional decision-making
- student dignity, privacy, and equity guide adoption decisions

A clear philosophy anchors policy decisions when new technologies emerge.

2. Use Categories

Many districts organize AI guidance into use categories that clarify appropriate contexts for tool use.

Category A – Permitted with Disclosure. Examples may include:

- brainstorming ideas
- language scaffolding
- draft feedback suggestions

These uses support thinking while maintaining student ownership of reasoning.

Category B – Restricted Use. Examples may include:

- generating first drafts of assignments
- completing independent assessments

These activities typically require clearer boundaries to preserve authentic student thinking.

Category C – Prohibited. Examples may include:

- submitting AI-generated work as personal reasoning
- uploading identifiable student data into unsecured platforms

Explicit categories reduce ambiguity and support consistent expectations across classrooms.

3. Oversight Structures

Effective AI governance typically includes structures that allow policies to evolve as technologies change. Examples may include an AI review committee, periodic tool evaluations, bias and equity audits, data privacy verification processes, and public transparency reporting

Governance becomes sustainable when oversight functions as a continuous system responsibility rather than a one-time approval process.

Replacement Risk Scale (System Version)

Purpose: Evaluate the degree to which an AI tool may replace, support, or interfere with student reasoning before adoption.

Artificial intelligence tools vary widely in how they influence learning processes. A replacement risk scale helps districts examine whether a tool primarily reduces administrative burden or whether it risks replacing the cognitive work that instruction intends to cultivate.

District teams may evaluate tools across a continuum of replacement risk.

1 – Minimal Risk: Reduces administrative or formatting burden; does not replace core reasoning process.

2 – Moderate Risk: May reduce some cognitive effort is used without structure.

3 – Significant Risk: Can replace meaningful cognitive work when unstructured.

4 – High Risk: Substitutes for central reasoning processes in learning tasks.

5 – Total Replacement: Automates the intended learning outcome entirely.

Tools identified at higher levels of replacement risk typically benefit from clearer guardrails or reconsideration before adoption.

AI Communication Guide for Boards & Families

Purpose: Provide clear and consistent messaging that helps school boards, families, and communities understand how artificial intelligence is used in district learning systems.

AI governance extends beyond internal policy. Districts often benefit from communicating clearly with boards and families about how artificial intelligence tools are integrated into teaching, learning, and administrative practice. When expectations are explained openly, communities are better positioned to understand both the opportunities and the safeguards associated with emerging technologies.

Board-Level Communication:

Communication with school boards typically emphasizes the district's guiding philosophy and oversight responsibilities. Messaging may highlight that:

- AI is increasingly present in society and education.
- Schools play a role in preparing students to use these tools responsibly.
- District policies prioritize human judgment rather than automated decision-making.
- Clear boundaries guide where AI tools may and may not be used.
- Data privacy protections and oversight structures remain active.

These themes help ensure that governance discussions remain anchored in educational purpose rather than technology trends.

Family Communication:

Families often benefit from clear explanations about how AI tools appear in classroom learning environments. Communication may address:

- how AI tools may support instruction or student learning
- situations where AI use is restricted or limited
- expectations for student transparency when AI tools are used
- ways families can discuss responsible AI use at home

When communication remains transparent and accessible, misunderstandings are less likely to develop and trust becomes easier to sustain. Clear communication helps prevent misinformation, while silence often allows uncertainty to grow.

AI Governance Review Process

Purpose: Provide a consistent framework for evaluating new AI tools before district adoption.

As artificial intelligence platforms evolve rapidly, districts benefit from review processes that examine new tools through multiple lenses before they are introduced into classrooms or administrative systems. A structured governance review helps leadership teams consider not only technical features but also instructional, ethical, and equity implications. When evaluation processes are consistent, districts can adopt tools thoughtfully rather than responding reactively to emerging technologies.

Review Criteria: Leadership teams may examine several dimensions when evaluating AI tools for district use:

- Cognitive replacement risk. Whether the tool supports learning processes or risks replacing core reasoning tasks.
- Bias and equity impact. Whether the system may reproduce cultural, linguistic, or demographic bias.
- Data privacy compliance. Whether the tool meets district, state, and federal data protection requirements.
- Alignment with instructional design goals. Whether the tool supports explanation, transfer, revision, and student reasoning.
- Transparency expectations. Whether users can understand how the tool operates and how outputs are generated.

When review criteria are established in advance, leadership teams can evaluate AI tools consistently across departments and initiatives. This approach helps ensure that technology adoption remains aligned with district learning goals and governance commitments.

Leadership Reflection

AI governance is not simply a technical or legal matter. It reflects the values a district chooses to embed in its learning system.

Artificial intelligence introduces powerful tools into educational environments. Governance determines whether those tools quietly reshape learning – or whether they are aligned intentionally with the purposes of schooling.

Leadership conversations therefore often return to several questions. Are we primarily reacting to risk, or are we designing systems deliberately? Are policies reinforcing fear of misuse, or cultivating responsible professional judgment? Are we managing technology – or shaping the culture in which it operates?

Thoughtful governance protects several commitments that define strong learning systems:

- student dignity
- professional judgment
- equity and access
- the development of human reasoning

Without governance, AI use can become fragmented and inconsistent. With overly rigid governance, experimentation and learning may narrow. Sustainable systems typically find balance between these extremes – establishing guardrails while preserving the conditions for innovation.

In an era shaped by intelligent tools, clarity becomes a central leadership responsibility. Governance communicates what a district values. When those values remain visible, technology is more likely to strengthen – rather than distort – the intellectual purposes of schooling.

Superintended Briefing: Establishing Ethical AI Governance

Artificial intelligence is entering school systems quickly, often through individual classroom experimentation or vendor partnerships. Without clear governance structures, districts may adopt tools that unintentionally reshape learning, decision-making, and student data practices.

For superintendents, AI governance is less about regulating technology itself and more about clarifying the educational values that technology must serve. When expectations remain unclear, AI adoption can become fragmented across classrooms and schools. Some educators may experiment freely, others may avoid tools entirely, and students may encounter inconsistent guidance about responsible use.

District leadership can therefore provide coherence by establishing shared principles for AI use across the system.

Several governance commitments typically guide effective AI integration:

Protect student dignity and data privacy. AI systems often rely on large data sets. District policies help ensure that student information is handled responsibly and that data collection practices remain transparent.

Preserve human judgment in educational decisions. AI tools can support planning, organization, and feedback, but professional decisions about instruction, assessment, and student support remain human responsibilities.

Audit tools for equity and bias. AI systems may unintentionally reproduce historical inequities if their training data reflects biased patterns. District review processes help ensure that technologies serve all students fairly.

Align AI use with learning goals. Technology adoption benefits from being evaluated through an instructional lens: Does the tool strengthen reasoning, accessibility, or feedback? Or does it replace the cognitive work students should perform themselves?

Communicate clearly with families and staff. Transparency builds trust. When districts explain how AI tools are used and how student information is protected, families and educators better understand the purpose and boundaries of technology integration.

Governance does not eliminate experimentation. Rather, it establishes the guardrails within which responsible innovation can occur.

The central leadership question becomes clear: Are our AI policies simply responding to new tools, or are they deliberately protecting the intellectual purposes of schooling?

System Leader Playbook: Governing AI with Integrity

AI governance shapes more than technology use. It influences the culture, expectations, and intellectual priorities of a district's learning system. When governance structures are thoughtfully designed, artificial intelligence can support reasoning, accessibility, and efficiency without displacing the human judgment that education depends on.

The following leadership moves illustrate how districts can establish governance structures that balance innovation with ethical responsibility.

1. **Establish an ethical AI review process.**

Districts benefit from reviewing AI tools through a structured evaluation process before adoption. Review criteria may include:

- cognitive replacement risk
- bias and equity implications
- student data protection
- alignment with instructional goals

Many districts use a structured protocol such as the Ethical AI Decision Compass introduced in Chapter 7 to guide these evaluations.

2. **Clarify transparency expectations.**

Clear expectations help students and educators understand when and how AI tools may be used responsibly.

Policies may clarify:

- when AI tools are permitted
- when use is restricted
- when disclosure is expected
- where decision authority remains human

Transparency helps reduce confusion while strengthening trust across the learning system.

3. **Protect human judgment.**

AI tools may assist with certain professional tasks, such as:

- drafting preliminary materials
- suggesting feedback language
- organizing information or documentation

However, core educational decisions typically remain human responsibilities. These may include grading determinations, instructional decisions, and individualized education planning.

Protecting human judgment ensures that technology supports – rather than replaces – professional expertise.

4. **Align AI use with learning design.**

AI integration is most productive when it aligns with the instructional design principles explored earlier in this book.

District guidance may encourage uses that:

- scaffold novice learners
- support language accessibility
- organize or visualize thinking

At the same time, learning systems often protect tasks where student reasoning remains essential. These may include:

- generating final explanations
- interpreting complex ideas
- engaging in productive intellectual struggle

When AI use aligns with learning design, technology strengthens rather than erodes cognitive development.

5. **Conduct periodic AI ethics reviews.**

Governance systems benefit from regular review as technologies evolve.

Periodic evaluations may examine:

- emerging bias patterns
- patterns of teacher and student use
- clarity of existing policies
- community questions or concerns

These reviews help ensure that governance remains responsive rather than static.

PART IV — SYSTEM ALIGNMENT

Chapter 9: Aligning Curriculum, Assessment, and AI

Closing the Gaps That Undermine Learning

Chapter 8 examined how districts can govern artificial intelligence ethically. Governance establishes important guardrails, yet policy alone rarely determines how technology ultimately influences learning. For AI integration to support reasoning rather than distort it, governance structures must align with the curriculum, assessment, and grading systems that shape everyday classroom incentives.

Research in the learning sciences consistently emphasizes the importance of alignment across instructional systems. Students develop durable understanding when learning goals, instruction, and assessment reinforce the same intellectual priorities. Simply put, knowledge becomes meaningful when it is "organized within conceptual frameworks that support transfer to new situations" (Bransford et al., 2000, p. 16).

In many districts, however, different parts of the system communicate different expectations about what learning matters most.

Students often receive mixed signals:

- Schools say they value transfer, yet many assessments emphasize recall.
- Schools emphasize thinking, yet grading systems often reward completion.
- Schools encourage depth, yet pacing structures frequently reward speed.

Artificial intelligence does not create these tensions. It reveals them more clearly. When systems are misaligned, AI tends to amplify the layer of the system that carries the strongest incentives.

These contradictions emerge because different structures within the learning system signal different priorities. Curriculum may emphasize conceptual understanding. Assessment may prioritize correct answers. Grading may reward completion and efficiency. AI tools then accelerate whichever layer of the system exerts the strongest pressure.

District systems rarely struggle because educators lack commitment or effort. More often, they struggle because the surrounding architecture sends conflicting signals. When curriculum, assessment, grading, and technology integration operate independently, coherence weakens and cognition fragments.

This chapter examines how these structural signals interact. By exploring the alignment among curriculum, assessment, grading, and AI integration, district leaders can begin to redesign systems so that each layer reinforces the same intellectual goals. When alignment strengthens, learning systems are more likely to cultivate reasoning, transfer, and durable understanding in an AI-rich environment.

Curriculum Signals

What the curriculum communicates about thinking

Curriculum communicates expectations about thinking long before students encounter formal assessment. The design of units, tasks, and learning experiences signals what forms of intellectual work the system values. These signals shape how students approach learning, how teachers structure instruction, and how effort is distributed across the classroom.

In some systems, curriculum structures emphasize procedural activity and completion. Units may prioritize coverage of topics, repetition of procedures, short-answer responses, and rapid task completion.

In other systems, curriculum design emphasizes deeper intellectual engagement. Learning experiences may prioritize:

- explanation of reasoning
- conceptual transfer across contexts
- sustained problem solving
- revision and iterative improvement

These design choices communicate powerful messages about what counts as learning. When curriculum structures reward completion alone, students often focus on efficiency. When curriculum structures invite explanation, transfer, and revision, students are more likely to engage in sustained reasoning.

As described in Part I of this book, learning environments that require explanation, transfer, and revision tend to cultivate reasoning rather than compliance. Curriculum architecture therefore becomes one of the earliest signals within a learning system about what kinds of thinking truly matter.

Assessment Signals

Assessment systems communicate powerful signals about what kinds of thinking count as success. Even when curriculum encourages explanation, interpretation, and transfer, assessments sometimes emphasize different priorities. When these signals diverge, students often adapt their effort toward what evaluation rewards.

In many learning systems, assessment structures tend to prioritize correctness of answers, speed of completion, and reproduction of memorized procedures.

At the same time, schools frequently describe learning goals in broader terms. Educational frameworks often emphasize:

- reasoning and explanation
- transfer of knowledge across contexts
- interpretation of complex ideas
- judgment under uncertainty

In AI-rich environments, these tensions become more visible. When assessments measure answers rather than reasoning, AI tools can often generate acceptable outputs without demonstrating meaningful understanding.

Standards Misaligned with Evaluation

District standards frequently articulate ambitious intellectual goals. These may include capabilities such as:

- analysis of complex ideas
- construction of arguments
- problem solving under constraint
- modeling real-world systems
- interpretation of evidence

Yet evaluation systems sometimes emphasize more procedural indicators of success, such as:

- accuracy of responses
- completion of assigned tasks
- formatting conventions
- timeliness of submission
- procedural compliance

When evaluation contradicts stated learning goals, students tend to follow incentives rather than philosophy. In this environment, polished output becomes the primary signal of success. AI tools can then become a logical strategy for producing that output.

Alignment improves when evaluation examines the capability named by the standard. For example, when standards emphasize argumentation, evaluation benefits from examining how students defend reasoning rather than simply how they structure written responses.

Evaluation Misaligned with Cognition

Evaluation systems often privilege what is easiest to quantify. Examples include multiple-choice accuracy, automated scoring systems, and rubric box-checking.

These measures provide useful information, yet they capture only part of the learning process. Cognition often involves additional capacities such as:

- transfer of knowledge to unfamiliar contexts
- adaptation of strategies when conditions change
- metacognitive reflection
- ethical reasoning about decisions
- selection among alternative approaches

When evaluation focuses primarily on easily measurable indicators, the intellectual scope of learning can narrow. AI tools can often complete surface-level tasks with efficiency.

When assessment systems invite students to explain, defend, and revise their thinking, ownership of reasoning becomes more visible. Alignment between standards, assessment, and cognition strengthens when evaluation examines not only what students produce, but how they arrive there.

What grading ultimately rewards

Grading systems often communicate the strongest signals about what behaviors matter in school. While curriculum and assessment describe learning goals, grading determines how those goals translate into incentives for students. The structures used to calculate grades can therefore influence how students allocate effort, interpret expectations, and approach learning tasks.

In many systems, grading practices tend to reward indicators such as completion of assigned work, speed of submission, procedural accuracy, and compliance with task requirements.

Alternative grading approaches emphasize different intellectual priorities. These systems often reward:

- explanation of reasoning
- revision and refinement of ideas
- defense of thinking
- transfer of knowledge to new contexts

The distinction becomes more visible in AI-rich environments. When grading emphasizes completion and polish, AI-generated responses can often meet expectations without demonstrating student ownership of reasoning. When grading emphasizes explanation, revision, and defense of thinking, the process of learning becomes more visible.

A Redesign Perspective

Grading systems sometimes weaken alignment within learning systems because they collapse multiple dimensions of learning into a single score. In many classrooms, grades may combine several skills into one numerical value, emphasize completion rather than development, discourage revision after evaluation, and close the learning cycle once a score is assigned.

In AI-rich environments, grading systems increasingly function best when they communicate information about learning rather than serve as a fixed judgment of ability. When grades signal progress, revision opportunities, and reasoning quality, they support the development of thinking rather than simply recording outcomes.

Redesign Principles

Several grading principles can help strengthen alignment between grading practices and learning goals.

1. **Separate Process from Product:** Evaluate reasoning and intellectual work separately from formatting or presentation.
2. **Normalize Revision:** Allow students to improve scores through documented refinement and learning.
3. **Weight Transfer Heavily:** Allocate meaningful value to applying knowledge in unfamiliar contexts.
4. **Include Defense Opportunities:** Provide moments where students explain and defend their reasoning.
5. **Provide Feedback Before Final Evaluation:** Offer formative feedback before assigning final scores so revision remains possible.

These structures help ensure that grading systems reinforce learning processes rather than prematurely concluding them.

System-Level Adjustments

District grading policies can also influence how grading signals operate across classrooms. Policies may clarify:

- expectations for AI disclosure in student work
- opportunities for iterative drafting and revision
- emphasis on conceptual mastery rather than task completion
- reduced weighting of tasks that can be easily automated
- protection of explanation and reasoning as core evaluation criteria

When grading structures reward thinking, students tend to prioritize thinking. When grading systems reward polished outputs alone, students often prioritize polish. Alignment strengthens when grading practices reinforce the intellectual goals that curriculum and assessment intend to cultivate.

AI Amplification Effects

How AI accelerates existing misalignment

Artificial intelligence rarely introduces entirely new problems into school systems. More often, it accelerates the incentives that already exist within curriculum, assessment, and grading structures. When those structures are aligned with reasoning and understanding, AI can support learning. When they emphasize speed, completion, or surface performance, AI tends to amplify those signals as well.

For example, when system incentives emphasize coverage, AI tools often accelerate summarization. When assessments prioritize correct answers, AI can generate acceptable responses quickly. When grading rewards completion and polish, AI-produced work may satisfy expectations with minimal cognitive ownership.

Conversely, when systems emphasize reasoning and intellectual development, AI tools may support learning processes such as brainstorming initial ideas, scaffolding language development, suggesting revisions during drafting, and accelerating feedback cycles.

In this sense, artificial intelligence functions less as a disruptive force and more as an amplifier. It strengthens the intellectual priorities that the surrounding system already communicates.

Assessment Misaligned with Transfer

Transfer frequently appears in district standards and learning frameworks, yet it is less often engineered into assessment structures. Many assessments continue to rely on formats that reward familiarity rather than adaptation.

Assessment structures sometimes include features such as:

- tasks that closely mirror practice examples
- predictable or familiar formats
- limited ambiguity in problem conditions
- minimal variation from rehearsed procedures

Students may perform successfully under repetition. When conditions shift, however, understanding may prove more fragile. AI systems are particularly effective at generating responses within familiar formats. Tasks requiring contextual interpretation, live explanation, or defense of reasoning remain more resistant to automation.

Alignment strengthens when evaluation measures the same capabilities that standards describe. If a standard emphasizes argumentation, assessment structures benefit from examining how students defend reasoning rather than simply whether responses resemble expected formats.

AI-Proof Assessment Audit

Assessment systems can also be examined systematically to determine whether they reveal student ownership of reasoning. In AI-rich environments, assessments increasingly function best when they require processes such as transfer, adaptation, and explanation rather than simply producing outputs.

District teams may examine major assessments by asking questions such as:

1. What cognitive capability does this assessment intend to measure? For example: recall, analysis, synthesis, or judgment.

2. Could an AI system generate an acceptable response without student reasoning? If so, the task may benefit from redesign.

   o Does the assessment invite processes such as explanation of reasoning, contextual variation, defense of ideas, revision opportunities, or live discussion or questioning?

3. Is success possible primarily through pattern replication? If so, the assessment may remain vulnerable to automation.

4. What signals does grading emphasize? When grading rewards structure or formatting more heavily than reasoning, incentives can shift toward polished output rather than intellectual ownership.

When assessments reveal how students explain, adapt, and defend their thinking, AI becomes less capable of substituting for the learning process itself. Alignment improves when evaluation structures illuminate reasoning rather than simply recording answers.

District Tools

System alignment rarely emerges through philosophical agreement alone. It develops when districts establish practical structures that allow leaders and educators to examine how curriculum, assessment, grading, and AI integration interact across the learning system.

The tools below help leadership teams identify where system signals diverge and redesign learning experiences so that reasoning, transfer, and ownership of thinking remain central.

These tools are commonly used by curriculum teams, instructional leaders, and professional learning communities engaged in system-level redesign.

Performance Task Conversion Lab

Purpose: Transform traditional assignments into transfer-oriented performance tasks that reveal student reasoning and reduce vulnerability to automation.

Many districts already possess strong standards-aligned tasks. In many cases, the challenge lies less in the content of those tasks and more in their format. The Conversion Lab model helps teams redesign existing assignments so that they require interpretation, judgment, and application rather than simple reproduction.

Step 1 – Identify the Core Standard

Clarify the conceptual understanding the task is intended to develop.

Questions to consider may include:

- What idea or principle should students understand?
- What intellectual capability does the standard describe?

Step 2 – Identify the Current Output

Examine the artifact students are currently asked to produce.

Examples may include an essay, a worksheet, a test response, a short written explanation.

This step helps teams see whether the task structure reveals reasoning or simply records output.

Step 3 – Identify Automatable Components

Review the task and identify elements that AI tools could easily reproduce.

Examples may include factual summaries, structured essay responses, procedural problem solving, or pattern-based explanations.

Recognizing these elements helps teams determine where reasoning may need to become more visible.

Step 4 – Add Transfer and Judgment Layers

Teams may then introduce additional conditions that invite interpretation and application.

Possible design layers include:

- contextual constraints
- role-based scenarios
- competing stakeholder perspectives
- counterargument defense
- live questioning or dialogue
- revision based on critique
- cross-disciplinary application

These additions shift the task from producing an artifact toward demonstrating ownership of reasoning.

Example Conversion

Traditional Task: Write an essay explaining the causes of a historical conflict.

Converted Performance Task: Advise a policy committee on how lessons from the historical conflict should inform a contemporary diplomatic decision. Present reasoning publicly and revise recommendations after peer counterarguments.

The structural shift moves the task from explanation to application, from recall to judgment, and from artifact production to reasoning ownership.

Coherence Mapping Across the System

Purpose: Examine whether cognitive expectations develop progressively across grade levels rather than resetting at each transition.

Alignment across grade bands helps ensure that reasoning skills deepen over time. When grade levels operate independently, students may encounter inconsistent expectations for explanation, revision, and transfer.

Vertical coherence mapping allows leadership teams to examine whether intellectual development progresses systematically.

Elementary Grades:

Review whether foundational reasoning structures appear early in the system. Questions to consider may include:

- Are explanation norms introduced during early learning experiences?
- Is revision visible as part of the learning process?

Middle Grades:

Examine whether learning tasks introduce increased cognitive complexity. Examples may include:

- exposure to ambiguity or variation
- opportunities to defend reasoning
- comparison of alternative strategies

These structures help students move beyond procedural replication.

Secondary Grades:

Evaluate whether learning tasks require sustained reasoning and transfer. Indicators may include:

- authentic or real-world problem contexts
- cross-disciplinary applications
- opportunities for live explanation and defense of reasoning

When cognitive expectations develop progressively across grade levels, students encounter increasing intellectual complexity rather than repeated procedural practice. System alignment supports a learning progression where explanation leads to transfer, and transfer leads to judgment.

The Graduate Profile for an AI Era

As artificial intelligence reshapes how information is produced and accessed, many districts are reconsidering the capabilities students will need beyond school. Defining a graduate profile can help clarify the forms of thinking and judgment that learning systems are designed to cultivate. When these capabilities are articulated clearly, curriculum, assessment, grading, and technology integration can align around shared intellectual priorities.

A graduate prepared for an AI-rich environment may demonstrate several interrelated capacities.

1. **Conceptual Understanding:** The ability to explain underlying principles and relationships rather than simply reproduce answers.
2. **Adaptive Reasoning:** The capacity to modify strategies when conditions change and unfamiliar problems emerge.
3. **Intellectual Humility:** A willingness to reconsider assumptions and revise beliefs in response to new evidence.
4. **Transfer Across Contexts:** The ability to apply knowledge and skills in unfamiliar situations or interdisciplinary contexts.
5. **Ethical Judgment:** The capacity to weigh consequences, perspectives, and values when making decisions.
6. **Cognitive Endurance:** The ability to sustain attention, effort, and reasoning through complex or ambiguous problems.
7. **Collaborative Thinking:** The ability to reason with others, engage in critique, and refine ideas through dialogue.

When districts articulate these capabilities explicitly, they provide a clearer reference point for instructional design and system alignment. Curriculum can emphasize conceptual understanding and transfer, assessments can examine reasoning under variation, and technology policies can support rather than substitute for the development of human judgment.

Leadership Reflection

System alignment rarely emerges through compliance checks alone. It develops when the structures of curriculum, assessment, grading, and technology reinforce the same intellectual priorities. Artificial intelligence did not create misalignment in school systems. In many cases, it has simply made existing contradictions easier to see.

When different parts of the system communicate different expectations, students tend to follow the strongest incentives available. If curriculum emphasizes reasoning but grading rewards completion, efficiency often becomes the dominant strategy. AI tools can accelerate that dynamic.

Alignment shifts the signal. When learning systems consistently emphasize capabilities such as judgment, transfer, explanation, revision, and ethical reasoning, ownership of thinking becomes more visible. In these environments, AI tools are less likely to replace learning processes and more likely to support them.

Leadership conversations therefore often move beyond the question of preventing misuse. A more generative question emerges: Have we designed systems that unmistakably reward human thinking?

Coherence strengthens learning systems. Alignment protects intellectual purpose. In an era shaped by intelligent tools, integrity becomes not only a philosophical commitment but also a matter of careful design.

Alignment is integrity. And in an AI era, integrity must be engineered.

Superintendent Briefing: Aligning Curriculum, Assessment, and AI

Artificial intelligence introduces new capabilities into learning environments, but its influence is often shaped by existing system incentives. When different elements of a district's instructional system communicate different priorities, students tend to follow the signals that appear most consequential. In many schools, those signals originate from grading policies, assessment structures, and accountability expectations.

For superintendents, the central challenge is therefore not simply managing AI use. It is ensuring that curriculum, assessment, grading, and technology all reinforce the same intellectual goals.

When alignment is weak, contradictions become visible. Curriculum may emphasize inquiry and reasoning, while assessments prioritize rapid completion of correct answers. Teachers may encourage revision and explanation, yet grading policies reward efficiency and final products. In such environments, AI tools can unintentionally amplify existing misalignment by making it easier to produce polished outputs without sustained reasoning.

District leadership can address this challenge by examining the signals embedded in system structures.

Curriculum alignment. Curriculum frameworks clarify the kinds of thinking students are expected to practice. District teams can examine whether units consistently invite explanation, interpretation, and transfer rather than primarily organizing information coverage.

Assessment alignment. Assessments benefit from capturing how students reason rather than only what answers they produce. Tasks that require contextual judgment, defense of thinking, or revision under feedback often reveal deeper understanding.

Grading alignment. Grading policies influence how students allocate effort. Systems that recognize revision, explanation, and conceptual understanding encourage intellectual ownership more consistently than systems focused primarily on completion.

Technology alignment. Artificial intelligence tools are most constructive when they support reasoning rather than replace it. District guidance can help ensure that AI use strengthens learning processes while protecting tasks where student thinking remains essential.

When these elements operate coherently, AI becomes less disruptive. Instead of competing with human learning, technology operates within structures designed to cultivate it.

The leadership question for district teams becomes clear: Do our systems consistently reward the development of thinkers, or do they still prioritize the production of answers?

Alignment ensures that the answer is unmistakable.

System Leader Playbook: Aligning the System Around Thinking

Instructional improvement rarely succeeds when different parts of a system communicate different expectations. Curriculum, assessment, grading, and accountability structures each send signals about what kinds of thinking matter. When these signals diverge, students often adapt their effort toward the incentives that appear strongest. System alignment helps ensure that each layer of the learning architecture reinforces the same intellectual priorities.

The leadership moves below illustrate ways districts can strengthen alignment around reasoning, transfer, and ownership of thinking.

1. **Examine assessments for AI vulnerability.**

District teams often benefit from reviewing major assessments to determine whether they reveal student reasoning or primarily capture finished outputs.

Guiding questions may include:

- Could an AI system generate an acceptable response independently?
- Does the task require explanation of reasoning?
- Does it invite transfer to unfamiliar contexts?
- Are students asked to defend or revise their thinking?

When these elements are absent, tasks may benefit from redesign so that student reasoning becomes more visible.

2. Convert traditional assignments into performance tasks.

Many traditional assignments can be strengthened by introducing elements that require interpretation, judgment, and application.

Assessment formats may incorporate features such as:

- contextual constraints
- role-based scenarios
- competing perspectives or counterarguments
- opportunities for live reasoning or discussion
- revision cycles following critique

These structures help shift assignments from artifact production toward demonstration of thinking.

3. Reconsider grading signals.

Grading systems communicate powerful incentives about what counts as success. Districts can explore grading structures that place greater emphasis on quality of reasoning, evidence of transfer, improvement through revision, and explanation of thinking.

At the same time, evaluation systems may reduce emphasis on outputs that can be easily automated.

4. **Map vertical cognitive progression.**

Alignment across grade levels helps ensure that reasoning develops progressively rather than restarting at each transition. Examples of developmental progression may include:

- Elementary grades: explaining ideas and reasoning clearly
- Middle grades: applying knowledge under variation and ambiguity
- Secondary grades: defending reasoning under critique or counterargument

Coherence across grade levels supports cumulative intellectual development.

5. **Align evaluation with intellectual priorities.**

Teacher observation and evaluation systems also communicate signals about instructional priorities. Some districts examine whether observation frameworks emphasize elements such as:

- cognitive architecture within lessons
- opportunities for student reasoning
- transfer across contexts
- visible revision culture

These indicators can complement more traditional measures such as pacing or engagement.

When curriculum, assessment, grading, and evaluation reinforce the same intellectual goals, alignment strengthens across the system. In these environments, instructional improvement becomes less dependent on isolated initiatives and more embedded within the architecture of the learning system.

Chapter 10: Remaining Accountability in the AI Era

Measuring What Actually Matters

Chapter 9 explored how districts can align curriculum, assessment, grading, and AI integration so that learning systems consistently reward reasoning. Alignment, however, introduces a related challenge: measurement. When accountability systems continue to focus primarily on visible outputs rather than cognitive development, misalignment can persist even within otherwise well-designed learning environments.

Many accountability systems were developed in an era when access to information was limited and producing finished work required significant effort. In that context, artifacts such as essays, problem sets, and written reports often served as reasonable indicators of student learning. Correct answers and completed assignments were frequently interpreted as evidence that thinking had occurred.

In an AI-rich environment, that relationship becomes less reliable. Digital tools can generate polished essays, solve structured problem sets, and draft reports with increasing fluency. The presence of finished work no longer guarantees that the underlying reasoning belongs to the student who submitted it.

For district leaders, the central question therefore begins to shift. The issue is not simply whether students are producing work. The deeper question is whether learning systems are cultivating judgment, interpretation, and intellectual ownership.

Accountability systems may therefore evolve from monitoring compliance with task completion toward examining the development of reasoning. When measurement focuses on cognitive capability rather than artifact production, accountability becomes more closely aligned with the intellectual purposes of schooling in an AI-enabled world.

What Are We Measuring?

Accountability systems communicate powerful signals about what educational systems ultimately value. The indicators chosen to monitor progress influence how districts allocate attention, how schools organize effort, and how educators interpret expectations. For this reason, the measures used within accountability frameworks shape not only reporting structures but also the incentives that guide daily practice.

Many existing accountability systems rely on indicators such as completion rates, pacing adherence, assessment proficiency levels, time-on-task observations, and alignment between lessons and stated objectives.

These measures can provide useful information about organizational efficiency and instructional consistency. However, they do not always reveal whether deeper cognitive development is occurring.

As districts consider accountability in an AI-rich environment, leadership conversations often return to a foundational question: What capabilities should graduates ultimately demonstrate?

When districts describe graduate outcomes, they frequently emphasize capacities such as:

- adaptive expertise in unfamiliar situations
- ethical reasoning about complex decisions
- transfer of knowledge across contexts
- sustained cognitive effort through challenging problems
- a sense of belonging and identity safety within learning environments

When these capabilities define the intended graduate profile, accountability systems benefit from examining how those constructs can be measured directly. Without this alignment, systems may unintentionally reward indicators that differ from the goals they publicly describe.

Accountability as Alignment

To connect measurement to educational purpose, accountability systems function most effectively when they reflect the intellectual purposes that districts intend to cultivate. In an AI-rich environment, remaining accountable does not require abandoning standards or expectations. Instead, it invites closer alignment between what systems measure and the capabilities they claim to value.

When measurement systems and learning goals diverge, the resulting signals can pull practice in unintended directions. Alignment improves when accountability indicators reflect the same intellectual priorities expressed in curriculum design, instructional practice, and graduate profiles.

Leadership teams may therefore examine questions such as:

- If the graduate profile emphasizes adaptive expertise, where is that capability visible in measurement systems?
- If ethical reasoning is a stated goal, how is it examined or documented in assessment structures?
- If sustained intellectual effort is valued, where do systems protect time for deep work?
- If belonging and identity safety are priorities, where are those experiences reflected in data or feedback systems?

These questions help districts examine whether accountability structures reinforce the goals they describe.

In aligned systems, measurement expands beyond procedural indicators to include constructs such as:

- reasoning and explanation
- transfer across contexts
- revision and intellectual growth
- belonging within learning environments
- judgment in complex situations

When accountability systems examine these dimensions, measurement becomes more closely connected to the intellectual and human purposes of education rather than functioning solely as a mechanism for compliance monitoring.

Measuring Adaptive Thinking

Adaptive thinking represents one of the central capabilities districts often hope to cultivate. It reflects a learner's ability to interpret unfamiliar situations, adjust strategies, and apply knowledge across contexts. Because these capacities unfold through reasoning rather than simple output, they may require different forms of observation and documentation than traditional performance indicators.

Adaptive thinking often becomes visible when students:

- modify strategies when conditions change
- explain why a particular method applies in a given situation
- compare alternative approaches to a problem
- revise reasoning in response to feedback
- apply knowledge in unfamiliar or interdisciplinary contexts

These behaviors can be observed and documented within learning environments, yet they are not always captured through traditional static performance indicators. Learning systems that seek to make adaptive thinking visible often examine evidence from sources such as:

- student dialogue and discussion
- iterative revision cycles
- performance-based assessments
- live reasoning or explanation
- application of knowledge across contexts

When measurement systems rely exclusively on static indicators of performance, they emphasize stability and replication. When evaluation structures make reasoning processes visible, learning environments are more likely to cultivate flexibility and judgment associated with adaptive expertise.

Protecting Deep Work

As artificial intelligence accelerates the production outputs, the distinction between surface-level output and deeper cognitive engagement becomes more important. Deep work refers to sustained intellectual effort that allows learners to grapple with complexity, refine ideas over time, and build durable understanding. This type of thinking often involves processes such as:

- sustained attention to challenging ideas
- tolerance for ambiguity and uncertainty
- productive cognitive struggle
- revision and refinement over multiple iterations
- defense of reasoning through discussion or questioning

Because these processes unfold gradually, learning systems sometimes benefit from protecting dedicated time and space for them within instructional schedules.

Accountability structures can support deep work by recognizing learning experiences that include opportunities for:

- device-free cognitive rehearsal
- early drafting and exploratory thinking
- slow reading and close interpretation
- multi-step reasoning processes
- structured peer critique and revision

These learning conditions may appear quieter or less visibly productive than rapid task completion. Yet they often represent moments when schema development and conceptual understanding are forming.

When systems expect constant visible productivity, opportunities for depth can narrow. Recognizing that sustained thinking sometimes appears quiet – or temporarily inefficient – helps accountability frameworks remain aligned with the cognitive processes that support meaningful learning.

Beyond Compliance Walkthroughs

Shift observation toward cognitive architecture. Classroom walkthroughs are often designed to provide quick snapshots of instructional practice. Many observation frameworks focus on visible indicators that signal organization, pacing, and alignment with instructional expectations. While these indicators can provide useful information about classroom structure, they may not always reveal the cognitive processes that learning environments are intended to cultivate.

Traditional walkthrough frameworks frequently emphasize indicators such as a clearly posted learning objective, visible student engagement, teacher circulation during activities, alignment between tasks and standards, and classroom management routines.

These elements help observers understand whether instructional structures are present. However, they do not necessarily illuminate how students are thinking or what intellectual work is occurring.

As learning systems adapt to an AI-rich environment, observation systems can expand their focus. In addition to noticing what is happening in a classroom, observers may also examine the kinds of thinking that instructional design invites.

Observation questions may therefore shift from: "What instructional actions are visible?" to questions such as: "What kinds of reasoning are students being asked to engage in?"

This shift helps align observation systems with the broader learning goals discussed throughout this book. When walkthroughs attend to cognitive architecture as well as visible structure, they provide richer insight into whether classroom environments are cultivating explanation, transfer, revision, and intellectual ownership.

Shifting accountability toward thinking requires more than philosophical agreement. District leaders benefit from practical tools that help make reasoning visible within observation systems and instructional monitoring processes. When observation structures focus only on visible activity, deeper cognitive processes may remain hidden.

The tools below help leadership teams examine whether classroom environments cultivate explanation, transfer, revision, and intellectual ownership rather than simply recording task completion.

Walkthrough 2.0: Thinking-Focused Observation Tool

Purpose: Provide an observation framework that examines the cognitive architecture of instruction rather than focusing only on visible activity.

Traditional walkthrough systems often emphasize visible classroom indicators such as engagement, pacing, or task completion. While these elements offer useful information, they do not always reveal the intellectual work students are being asked to perform. A thinking-focused observation tool expands observation domains so that leaders can examine whether instructional design invites reasoning, explanation, and revision.

Observation domains may include several dimensions of cognitive engagement.

Domain 1 – Cognitive Tension

Examine whether learning tasks introduce problems that require reasoning.

Indicators may include:

- students generating ideas before receiving explanation
- problems that require interpretation rather than simple recall

Domain 2 – Explanation and Defense

Observe whether students are asked to articulate and justify their reasoning.

Examples may include:

- students explaining how they arrived at a conclusion
- competing perspectives or counterarguments being discussed

Domain 3 – Revision Culture

Consider whether learning environments make intellectual refinement visible.

Indicators may include:

- students revising work after feedback
- iterative improvement of ideas or explanations

Domain 4 – Transfer

Examine whether students apply knowledge beyond familiar formats.

Examples may include:

- applying concepts in unfamiliar contexts
- selecting strategies under variation or ambiguity

Domain 5 – Risk Survivability

Observe whether the learning environment supports intellectual risk-taking.

Indicators may include:

- mistakes treated as sources of information
- students participating without fear of embarrassment or penalty

In this model, observers record evidence of thinking, not simply evidence of activity. The walkthrough becomes a diagnostic instrument that helps leaders understand how instructional design supports reasoning.

Cognitive Depth Index

Purpose: Provide a system-level indicator that helps districts examine the distribution of cognitive demand across classrooms and schools.

The Cognitive Depth Index (CDI) offers a structured way to examine instructional rigor across a district. Rather than evaluating individual lessons in isolation, the CDI helps leadership teams observe patterns in how cognitive demand develops across classrooms and grade levels.

Classrooms may be examined along a continuum of cognitive depth.

Level 1 – Recall Dominant

- reproduction of information
- minimal explanation of reasoning
- tasks limited to familiar formats

Level 2 – Explanation Emerging

- students explain reasoning processes
- structured opportunities for productive struggle

Level 3 – Transfer Visible

- variation in problem contexts
- students selecting among strategies
- adaptation required to complete tasks

Level 4 – Adaptive Judgment

- defense of reasoning under critique or counterargument
- interdisciplinary applications
- revision following feedback or critique

District leaders may use these observations to examine patterns such as:

- the distribution of cognitive depth across schools
- vertical coherence across grade levels
- progression of cognitive demand over time

The purpose of the index is not uniformity across classrooms. Instead, it helps districts observe patterns of growth as learning systems increasingly emphasize explanation, transfer, and adaptive reasoning. The goal is not uniformity – but growth.

Identity & Belonging Metrics

Purpose: Examine whether learning environments support the psychological conditions that allow students to engage in intellectual risk-taking and sustained reasoning.

Cognitive engagement rarely develops in isolation from students' sense of identity and belonging within a learning environment. When students feel psychologically safe to contribute ideas, challenge assumptions, and revise their thinking, deeper reasoning becomes more likely. When belonging is fragile, participation and intellectual risk-taking often narrow.

For this reason, some districts complement traditional accountability indicators with measures that examine whether learning environments support inclusion, voice, and intellectual agency.

Belonging-focused metrics may examine conditions such as:

- whether students feel safe sharing ideas during discussion
- whether student voice is distributed across demographic groups
- whether revision is encouraged without penalty or stigma
- whether instructional materials reflect diverse identities and perspectives
- whether students report a sense of intellectual agency in learning

These indicators help illuminate whether learning environments support the relational conditions that allow cognition to flourish.

When accountability systems incorporate belonging metrics, they begin to align measurement not only with academic performance but also with the conditions that support intellectual formation. In this sense, belonging functions not as a peripheral measure but as a foundational element of learning architecture.

Leadership Reflection

Artificial intelligence often reveals the underlying signals within accountability systems. When measurement focuses primarily on artifacts, systems tend to produce more artifacts. When evaluation attends to reasoning and judgment, the signals shift toward deeper thinking.

Accountability itself is unlikely to disappear. Educational systems will continue to examine evidence of learning, progress, and instructional effectiveness. The more significant question concerns what those systems choose to measure.

Leadership conversations increasingly return to a central consideration: accountable to what? To pacing schedules, production of assignments, and visible activity? Or to capacities such as human judgment, sustained cognitive effort, adaptive reasoning, and ethical decision-making?

The presence of AI tools does not eliminate accountability. Instead, it clarifies the distinction between outputs that can be automated and capabilities that remain distinctly human.

When measurement systems focus on reasoning, interpretation, and intellectual ownership, accountability aligns more closely with the purposes of education. In this sense, leadership involves protecting the conditions in which those human capacities can continue to develop.

Superintendent Briefing: Accountability in an AI Era

Artificial intelligence has revealed an emerging challenge within many accountability systems. For decades, districts have relied on visible artifacts – completed assignments, correct answers, and polished products – as indicators of learning. In an AI-rich environment, those artifacts no longer function as reliable evidence of student thinking.

When digital systems can generate essays, solve structured problems, and produce fluent reports with increasing ease, accountability frameworks may benefit from shifting attention toward cognitive processes rather than finished outputs. The central question becomes less about whether work is produced and more about whether reasoning is visible.

District leadership conversations therefore often focus on indicators such as:

- whether students explain the reasoning behind their conclusions
- whether knowledge is applied in unfamiliar contexts
- whether ideas are defended under critique or questioning
- whether thinking is revised in response to feedback

These indicators more closely reflect the capabilities many districts describe within their graduate profiles – capabilities such as adaptive expertise, ethical reasoning, transfer across contexts, and sustained cognitive effort.

Reframing accountability does not mean abandoning expectations or standards. Instead, it invites closer alignment between what districts say they value and what measurement systems actually examine. When accountability focuses primarily on compliance indicators, compliance tends to increase. When measurement attends to reasoning and intellectual ownership, learning systems begin to emphasize those capabilities.

In an AI-enabled environment, accountability systems increasingly serve a protective role: helping ensure that the development of human judgment remains central to the purposes of education.

System Leader Playbook: Designing Accountability for Thinking

Accountability systems shape how educators interpret priorities within a learning system. When measurement emphasizes completion, speed, or procedural compliance, those signals often influence instructional practice. When accountability frameworks attend to reasoning, transfer, and intellectual development, they help reinforce the cognitive goals districts describe in their graduate profiles.

The leadership moves below illustrate ways districts can begin aligning accountability systems with the development of thinking.

1. **Redesign walkthrough protocols.**

Observation systems can expand beyond compliance indicators such as posted objectives or visible engagement. Walkthroughs may examine evidence of learning processes such as:

- student reasoning and explanation
- productive intellectual struggle
- revision and refinement of ideas
- transfer of knowledge across contexts
- defense of ideas during discussion

In this approach, walkthroughs function less as compliance checks and more as diagnostic tools for understanding cognitive architecture within classrooms.

2. **Track cognitive depth across the system.**

District leaders may also benefit from examining patterns of cognitive demand across classrooms and schools. Tools such as the Cognitive Depth Index can help leadership teams observe:

- the distribution of recall-oriented versus reasoning-oriented tasks
- the presence of transfer opportunities within instruction
- the progression of cognitive demand across grade levels

The goal is not uniformity across classrooms, but a system-wide increase in intellectual depth over time.

3. **Protect deep work.**

Accountability frameworks can also recognize that meaningful learning often requires sustained time for thinking. Learning environments may benefit from protecting opportunities for:

- extended reasoning and problem solving
- iterative revision cycles
- peer critique and collaborative reflection
- engagement with ambiguity and productive struggle

When systems prioritize constant visible productivity, opportunities for depth can narrow. Leadership attention to these conditions helps preserve spaces where deeper cognition develops.

4. Measure belonging as a cognitive condition.

Students are more likely to take intellectual risks when they feel psychologically safe to share developing ideas. Some districts therefore examine indicators such as:

- student perceptions of safety when sharing ideas
- distribution of voice during discussions
- representation of identity and perspective within curriculum materials
- opportunities for revision without penalty

These indicators help illuminate the relational conditions that support cognitive engagement.

5. Align accountability with the graduate profile.

Accountability systems become more coherent when they reflect the capabilities districts hope graduates will demonstrate. Graduate profiles often emphasize capacities such as:

- adaptive expertise in unfamiliar situations
- ethical reasoning about complex decisions
- transfer of knowledge across contexts
- intellectual humility when revising ideas

When accountability structures examine these capabilities directly, measurement aligns more closely with the purposes of learning systems. Without that alignment, systems may unintentionally incentivize different outcomes than those they publicly describe.

Chapter 11: Protecting Teacher Energy & Sustainability

Designing Systems That Preserve Professional Capacity

Earlier chapters in this book examined the cognitive architecture that supports student thinking. This chapter turns to another dimension of learning systems: the professional capacity required to sustain those environments over time. Instructional redesign cannot endure if the educators responsible for it operate under conditions of chronic overload.

Teacher energy is therefore not simply a matter of morale. It is a structural component of system design. When professional attention is fragmented by competing initiatives, excessive monitoring, or expanding tool ecosystems, instructional focus can become difficult to sustain.

Artificial intelligence introduces a paradox within this landscape. Many technologies promise efficiency, automation, and relief from administrative burden. Yet in some districts the rapid introduction of new tools has produced the opposite effect: more platforms to manage, more decisions to navigate, more expectations to meet, and more cognitive noise within the workday.

In this environment, AI can function in two very different ways. It may help reduce extraneous workload and preserve professional attention for teaching. Or it may layer additional complexity onto already demanding systems.

The difference rarely lies in the technology itself. It lies in how systems are designed.

This chapter examines how districts can structure AI integration, professional expectations, and organizational routines in ways that protect educator energy while preserving the intellectual priorities described throughout this book. When professional capacity is sustained, instructional improvement becomes far more likely to endure.

AI as Amplifier of Burnout — or Reducer?

AI tools do not automatically increase or decrease teacher workload. Their impact depends largely on how they are integrated into existing instructional systems. When technology removes routine friction from daily work, professional capacity can expand. When new tools introduce additional expectations or complexity, the opposite effect may occur.

In some contexts, AI tools support teachers by reducing routine administrative burden. For example, they may assist with tasks such as drafting routine communications, summarizing instructional data, generating practice variations for students, organizing planning templates, and accelerating feedback cycles. Used in this way, AI can help preserve professional attention for the instructional decisions that require human judgment.

At the same time, technology integration can sometimes introduce new sources of strain. Burnout pressures may increase when systems:

- add new platforms without retiring existing ones
- expand expectations for personalization beyond sustainable levels
- increase documentation or reporting requirements
- require constant evaluation of emerging tools
- create uncertainty about policy or expectations

In these situations, the challenge often lies less in the tool itself and more in the accumulation of decisions surrounding its use. Professional energy tends to erode through friction, ambiguity, and cognitive overload rather than through any single demand.

Leadership conversations therefore often return to a simple question: Does AI integration reduce unnecessary cognitive load for teachers, or does it add new layers of complexity? The answer frequently depends on how intentionally the surrounding system is designed.

Decision Fatigue in an AI-Rich Environment

As artificial intelligence tools become more common in classrooms, teachers often encounter an expanding number of small instructional decisions throughout the day. Many of these choices involve determining how and when technology should be used within learning tasks. While each decision may appear minor on its own, the cumulative effect can place additional demands on professional attention.

Teachers may find themselves navigating questions such as: Should AI tools be allowed for this task? Should this activity be device-open or device-limited? Does this draft reflect student-authored thinking? How much confidence should be placed in this output? Is this tool approved within district guidance?

When these questions repeat across multiple classes, large numbers of students, and daily instructional planning, the volume of decisions can grow quickly. Over time, sustained decision pressure can affect qualities that educators rely on for effective teaching, including clarity in instructional judgment, patience during complex learning moments, consistency across classroom expectations, and creativity in lesson design.

Learning systems that seek to preserve teacher energy often attempt to reduce unnecessary decision burden. This can occur through structures such as:

- clear and accessible policy guidance
- shared design language across teams
- defined boundaries for technology use
- consistent expectations across classrooms and schools

When expectations remain ambiguous, teachers may spend significant energy interpreting rules rather than focusing on learning design. In this sense, uncertainty often drains professional capacity more quickly than workload alone.

Tool Overload

Technology can fragment professional attention. As digital platforms expand across school systems, teachers often navigate an increasing number of instructional tools during the course of a typical workday. Each system may be introduced with the goal of improving efficiency, communication, or instructional insight. Over time, however, the accumulation of tools can create new layers of complexity within daily practice.

In many districts, educators interact with multiple platforms such as learning management systems, AI-supported writing or planning tools, grading and feedback platforms, assessment dashboards, communication applications, and data visualization systems.

Individually, each tool may offer useful capabilities. When these systems accumulate without careful coordination, however, teachers may find themselves moving constantly between platforms in order to complete routine tasks.

This environment can contribute to several pressures within professional practice:

- fragmented attention across multiple systems
- increased time spent navigating interfaces rather than designing instruction
- reduced space for reflective judgment within teaching decisions

Technology adoption can therefore create unintended complexity when systems expand faster than they are streamlined. In these situations, innovation may appear as accumulation rather than coherence.

Sustainable system design often involves thoughtful subtraction as well as addition. When districts examine how tools interact within the broader instructional architecture, they are better positioned to preserve professional attention for the work of teaching itself.

Professional Clarity as Protection

Clear priorities preserve teacher energy. Professional energy is often sustained when educators understand which priorities matter most within a learning system. Clear expectations help teachers allocate attention, make instructional decisions with confidence, and avoid unnecessary pressure from competing initiatives. In this sense, clarity functions as a protective boundary around professional work.

Teachers frequently benefit from shared understanding about questions such as:

- which instructional priorities should receive the greatest attention
- which expectations are optional or flexible
- which elements of practice are protected within the system

When priorities remain unclear, educators may feel pressure to respond to every request or initiative simultaneously. Over time, this reactive urgency can drain professional capacity.

Learning systems that communicate priorities clearly allow teachers to make more deliberate choices. For example, when educators understand that:

- explanation carries greater importance than speed
- transfer of knowledge matters more than surface polish
- revision is valued more than immediate perfection

they are better able to allocate time and attention toward the intellectual work that matters most.

Clarity can therefore stabilize professional practice. When systems articulate a small set of non-negotiable priorities and release less essential demands, teachers gain greater space to focus on meaningful learning design.

District Tools

Protecting teacher energy requires more than goodwill. Sustainable systems rely on structures that help leaders examine how professional time and cognitive attention are distributed across the workday. When workload accumulates through fragmented tools, unclear expectations, and redundant processes, even highly committed educators can experience sustained strain.

The tools below help district teams identify where professional energy is depleted, where it is sustained, and how system design can better protect the intellectual work of teaching. By mapping energy patterns, simplifying tool ecosystems, and clarifying expectations around AI integration, leaders can redesign systems so that professional attention remains focused on learning rather than unnecessary complexity.

Teacher Energy Mapping Tool

Purpose: Help districts identify where professional energy is depleted or sustained so that systems can reduce unnecessary workload and protect instructional capacity.

Teacher workload is not experienced evenly across all tasks. Some professional activities support intellectual engagement and collaboration, while others consume time without strengthening instructional practice. Mapping how energy flows through a typical workweek can help leadership teams identify where system structures may be draining professional capacity.

The goal of energy mapping is not to evaluate individual resilience, but to surface patterns within the system that influence teacher workload and sustainability.

Step 1 – Map Weekly Energy Flow

Teachers or teams begin by examining recurring professional responsibilities and estimating how each activity affects professional energy.

For each task, educators may rate its impact as (1) High Energy Gain, (2) Neutral, (3) Moderate Drain, or (4) Severe Drain.

Common task categories may include lesson planning, grading and feedback, IEP documentation, parent communication, professional learning community meetings, experimentation with new tools or AI supports, and administrative compliance requirements.

When these ratings are examined across teams or departments, patterns often begin to emerge.

Step 2 – Identify Structural Energy Drains

After mapping energy patterns, leadership teams can examine which drains originate from system structures rather than individual preferences.

Questions for reflection may include:

- Is this task essential to learning, or does it persist because of legacy structures?
- Could AI tools reduce routine or administrative components of the task?
- Could policy changes eliminate redundancy across systems?

These conversations help shift the focus from personal workload management toward systemic redesign.

Step 3 – Protect High-Value Energy Zones

Energy mapping can also reveal professional activities that sustain meaningful instructional work. These areas often represent moments when teachers engage most deeply with student thinking. Examples may include:

- extended planning blocks focused on lesson design
- revision conversations with students
- classroom discussions centered on reasoning
- collaborative analysis of student thinking

Protecting these moments within schedules and policies helps preserve the intellectual core of teaching.

Energy invested in thinking often yields long-term instructional benefits. Energy consumed by fragmentation, by contrast, can contribute to professional exhaustion. Mapping these patterns allows districts to redesign systems so that professional attention remains focused on learning rather than on unnecessary complexity.

AI Overload Reduction Plan

Purpose: Help districts reduce unnecessary technological complexity so that AI integration preserves professional attention rather than fragmenting it.

As new digital platforms and AI tools enter school systems, districts may find that the number of required tools grows faster than existing ones are streamlined or removed. While each platform may promise efficiency, the cumulative effect can increase cognitive fragmentation for educators. A structured reduction plan helps leadership teams examine how technology systems interact and identify opportunities to simplify the professional environment.

The goal is not to limit innovation but to ensure that technology integration strengthens instructional systems rather than creating unnecessary complexity.

1. Tool Consolidation Audit.

Leadership teams may begin by reviewing the district's current digital ecosystem. An inventory process can examine questions such as:

- Is this tool essential to instructional or operational work?
- Does it duplicate functions available in another platform?
- Could the same purpose be achieved with fewer systems?

Many districts find value in periodically retiring or consolidating platforms as new tools are adopted, helping maintain coherence across the system.

2. **AI Boundary Clarity.**

Teachers often face uncertainty about when and how AI tools should be used in instructional contexts. Clear guidance can reduce decision fatigue. District policies may therefore clarify:

- where AI use is encouraged
- where AI use remains optional
- where AI use is restricted

When boundaries remain visible and consistent, educators can make instructional decisions with greater confidence.

3. **Documentation Streamlining.**

AI tools can assist with drafting reports, summaries, and other administrative documentation. However, efficiency gains may diminish if documentation expectations expand simultaneously. Leadership teams may therefore examine whether:

- AI tools reduce routine drafting time
- reporting requirements remain focused on essential information
- duplicate documentation processes can be eliminated

The goal is to ensure that technological efficiency supports professional capacity rather than generating additional reporting demands.

4. Protected Planning Windows.

Professional schedules play an important role in sustaining instructional quality. Some districts design schedules that protect time for activities such as:

- device-limited planning sessions
- collaborative design labs with colleagues
- reduced meeting frequency when possible

Protecting these windows helps ensure that professional attention remains available for lesson design, reflection, and collaboration.

When technology ecosystems remain coherent and expectations are clearly defined, AI tools are more likely to reduce routine friction rather than adding new layers of complexity. In this way, system design helps preserve the professional energy required for meaningful teaching and learning.

Strategic Simplification Framework

Purpose: Help district leaders evaluate initiatives, tools, and expectations so that system design protects teacher capacity and focuses attention on the intellectual work of teaching.

School systems often accumulate initiatives, tools, and reporting structures over time. While many of these efforts begin with positive intentions, their combined effect can fragment professional attention and dilute instructional focus. Strategic simplification offers a structured way for leadership teams to examine whether system demands support or compete with core learning priorities.

Rather than removing expectations indiscriminately, simplification encourages districts to apply consistent filters when evaluating new initiatives or existing requirements.

Filter 1 – Does This Strengthen Judgment?

District leaders may examine whether a policy, tool, or initiative contributes meaningfully to professional or student reasoning. Questions for reflection may include:

- Does this practice help educators or students develop stronger judgment?
- Does it support the intellectual goals described in the district's graduate profile?

If the connection to reasoning remains unclear, teams may reconsider its priority within the system.

Filter 2 – Does This Reduce Cognitive Noise?

Some initiatives improve clarity, while others add additional layers of complexity. Leadership teams may ask:

- Does this tool streamline existing work or introduce additional fragmentation?
- Does it simplify decision-making for educators?

When new systems increase cognitive noise, redesign or consolidation may be beneficial.

Filter 3 – Does This Protect Teacher Energy?

Professional capacity remains one of the most important resources within a learning system. Leaders may therefore examine:

- whether a requirement consumes disproportionate energy relative to its impact
- whether components of the task could be eliminated or automated responsibly
- whether expectations can be simplified while preserving their intended purpose

Strategic simplification does not represent reduced ambition. Instead, it reflects disciplined system design. When districts focus attention on what matters most, professional energy is more likely to remain available for the intellectual work of teaching and learning.

Leadership Reflection

Artificial intelligence does not determine whether teachers experience sustainable workloads. The surrounding system often shapes that outcome. When new tools are layered onto already complex structures, professional strain may increase. When technology removes routine friction and clarifies priorities, teacher capacity can expand.

The central question for leadership therefore becomes less about the tool itself and more about how it is integrated within the system. Are new technologies simplifying professional work or adding additional layers of expectation? Do educators have clarity about what matters most, or are competing initiatives pulling attention in multiple directions?

Professional energy, like student attention, is finite. Sustained teaching requires time, focus, and cognitive space. Systems that protect these conditions tend to support long-term instructional improvement, while systems that fragment attention often make meaningful work more difficult to sustain.

For this reason, the architecture of professional work matters as much as the architecture of learning. Instructional systems rely not only on sound design for students but also on structures that preserve teacher energy and judgment.

Professional endurance is therefore not simply a matter of individual resilience. It reflects the conditions created by system design.

Superintendent Briefing: Protecting Teacher Capacity

AI integration can strengthen instructional systems when it preserves the professional capacity of educators. Many technologies promise efficiency, yet without careful system design they can introduce new layers of complexity – additional platforms, more decisions, and greater cognitive noise within the workday.

Teacher energy functions as a finite resource within any learning system. When new tools expand monitoring requirements, documentation, or personalization expectations without reducing existing demands, workload may increase rather than decrease.

For this reason, many districts begin to treat professional capacity as a design variable rather than an incidental outcome. AI tools can be integrated in ways that remove routine friction and allow educators to focus on the work that requires human judgment: diagnosing student understanding, guiding inquiry, and cultivating reasoning.

Sustainable integration therefore often involves selective simplification as well as innovation. When new tools or expectations are introduced, systems may also examine which processes can be streamlined, consolidated, or retired.

When AI reduces unnecessary friction, professional capacity can expand. When it multiplies demands, the opposite effect may emerge. The difference frequently lies in how intentionally the surrounding system is designed.

System Leader Playbook: Designing Sustainable Teacher Work

Sustainable instructional improvement depends not only on strong curriculum and assessment design, but also on the conditions that support professional work. When system expectations accumulate without coordination, teacher attention can fragment and instructional focus may narrow. Leadership teams therefore often examine how policies, technologies, and routines interact so that professional capacity remains available for the work that requires human judgment.

The leadership practices below illustrate ways districts can design systems that protect educator energy while supporting thoughtful AI integration.

1. **Automate Administrative Friction.** Remove before adding. When new tools or initiatives are introduced, leadership teams may also examine which existing processes could be consolidated or retired. Without periodic subtraction, complexity can grow as new requirements accumulate.

2. **Protect Cognitive Work.** AI tools can assist with tasks that rely primarily on routine organization rather than professional interpretation. Examples may include:

    - drafting routine communications
    - summarizing data reports
    - organizing planning templates
    - formatting documentation

In many districts, automation is most effective when it targets clerical workload rather than instructional decision-making.

3. **Simplify Platform Ecosystems.** Teaching involves sustained intellectual activity that benefits from protected time. Educators often require space for work such as:
    - lesson design and refinement
    - analyzing student thinking
    - collaborative planning with colleagues
    - providing meaningful feedback

AI integration can be most helpful when it expands the time available for these activities rather than reducing it.

4. **Simplify Platform Ecosystems.** Over time, digital platforms can accumulate across instructional systems. Leadership teams sometimes review the ecosystem by examining:
    - the number of required tools in daily use
    - overlapping platforms that serve similar functions
    - redundant reporting or documentation systems

Streamlined platforms can help reduce cognitive noise and simplify daily workflows.

5. **Align AI with Instructional Design.** AI tools may support instructional architecture when they assist teachers in organizing learning experiences. In some cases, they may help with tasks such as:

   - generating practice variations for students
   - supporting scaffolds for multilingual learners
   - organizing feedback and revision cycles

At the same time, professional judgment remains central to teaching. AI tools are most effective when they support – rather than replace – the reasoning and decision-making of educators.

PART V — LEADERSHIP IMPLEMENTATION

Chapter 12: 90-Day District Redesign Plan

Moving from Philosophy to Practice

The preceding chapters explored the foundations of instructional redesign: teaching as design, adaptive expertise, ethical AI governance, system alignment, and the protection of teacher capacity. Together, these ideas outline a framework for learning systems that cultivate reasoning, judgment, and sustainable professional practice.

Yet ideas alone rarely transform districts. Change becomes visible when philosophy is translated into structures that guide everyday work. Professional learning, leadership conversations, and policy documents may introduce new directions, but sustained improvement typically depends on routines that support reflection, testing, and refinement over time.

District transformation therefore rarely emerges from isolated enthusiasm, single workshops, or one-time initiatives. While these moments may spark interest, system change more often develops through disciplined cycles that allow teams to examine evidence, adjust designs, and build coherence across classrooms and schools.

This chapter introduces a practical implementation structure designed to support that process. The 90-day Redesign Cycle provides a manageable timeframe for district teams to test ideas, observe their impact, and refine learning architecture in iterative stages.

The goal is not immediate perfection. The goal is momentum – steady movement from philosophy toward practice, supported by cycles of design, observation, and adjustment that gradually strengthen the system as a whole.

The 90-Day Redesign Architecture

Ideas introduced in earlier chapters become meaningful when they translate into routines that guide system improvement. Many districts find that structured redesign cycles help move from philosophical alignment toward practical implementation. A 90-day cycle offers one way to organize this work, providing enough time to examine evidence, test adjustments, and refine structures without losing momentum.

Within this model, district teams typically move through four phases:

- Diagnose — Where does misalignment appear within the system?
- Pilot — Which design adjustments show promise in practice?
- Iterate — What evidence suggests refinement is needed?
- Scale — Which structures are ready to expand across the system?

This cycle helps balance planning and action. It encourages experimentation without bypassing reflection and supports expansion only after evidence begins to accumulate. Ninety days often proves long enough to generate meaningful insight while short enough to maintain focus.

Phase 1: Diagnose (Weeks 1–3)

Purpose: Examine system architecture to identify areas of misalignment before redesign begins.

Effective diagnosis benefits from multiple forms of evidence rather than isolated anecdotes.

Step 1 – Coherence Snapshot

District teams may review indicators such as:

- curriculum alignment across grade levels
- assessment vulnerability to AI-generated responses
- clarity and consistency in grading practices
- distribution of cognitive depth across classrooms
- consistency of device and technology policies

Tools introduced earlier in this book – such as the AI-Proof Assessment Audit, Cognitive Depth Index, and Instructional Coherence Audit – can support this analysis. The goal is to document patterns where system signals may diverge.

Step 2 – Energy and Sustainability Scan

Teacher perspectives provide important insight into how system design operates in daily practice. Leadership teams may ask questions such as:

- Where does professional overload appear most frequently?
- Where is AI integration creating uncertainty or friction?
- Where is sustained cognitive depth most difficult to maintain?

These conversations help ensure that redesign efforts account for teacher capacity as well as instructional goals.

Step 3 – Identify One Redesign Target

Rather than attempting broad systemic overhaul, many districts begin with a focused design priority. Possible targets might include:

- redesigning transfer-oriented tasks
- implementing a thinking-focused walkthrough model
- clarifying AI governance policies
- strengthening revision culture within classrooms
- redesigning PLCs as collaborative design studios

Concentrated focus often increases the likelihood that meaningful learning will occur during the cycle.

Phase 2: Pilot (Weeks 4–7)

Purpose: Test the redesign in a contained environment before expanding system-wide.

Pilots allow leadership teams to examine how new structures operate under real classroom conditions. Pilot sites may include one school, one department, one grade band, or one leadership team.

During the pilot, teams often clarify several elements:

1. the capability outcome the redesign intends to strengthen
2. the cognitive architecture embedded in the learning design
3. the AI guardrails that support ethical integration
4. the evidence sources that will be examined during reflection

Evidence may include:

- samples of student reasoning
- observation notes from classrooms
- teacher reflection logs
- indicators of professional energy or workload impact

Pilots function as learning environments for the system itself. Unexpected outcomes often provide valuable information for redesign.

Phase 3: Iterate (Weeks 8–10)

Purpose: Refine the instructional architecture based on evidence gathered during the pilot.

During this phase, leadership teams examine how the design functioned in practice. Adjustments may include:

- recalibrating cognitive tension within tasks
- strengthening opportunities for revision
- clarifying boundaries for AI use
- reducing sources of unnecessary friction

Reflection questions may include:

- Did opportunities for adaptive thinking increase?
- Did classroom environments remain psychologically safe?
- Did teacher energy improve or decline during the pilot?
- Did AI tools support thinking or unintentionally replace it?

Iteration helps ensure that learning from the pilot informs subsequent decisions before expansion occurs.

Phase 4: Scale (Weeks 11–12)

Purpose: Extend structures that demonstrate promise across additional parts of the system.

Scaling typically becomes more effective when certain conditions are present:

- clear documentation of the redesigned structures
- processes that can be replicated by other teams
- coaching practices aligned with the new design
- policy or communication structures that reinforce expectations

At the same time, leadership teams often avoid expanding prototypes that remain ambiguous or unrefined. Scaling tends to amplify existing patterns, whether strengths or weaknesses. Ensuring architectural stability before expansion helps support sustainable system growth.

Design Studio Agendas

Structure collaborative redesign conversations. As districts move through redesign cycles, the structure of collaborative meetings can influence whether conversations remain focused on instructional architecture or drift toward routine updates. Many leadership teams therefore benefit from using structured studio agendas that center learning design, evidence of student thinking, and practical next steps.

Design studios function best when the agenda keeps attention on how lessons cultivate reasoning, transfer, and revision rather than on procedural reporting.

A 90-minute studio agenda may include:

1. Recenter on the Human Capability Goal (10 minutes). Briefly revisit the intellectual capability the redesign is intended to strengthen.

2. Present Pilot Architecture (15 minutes). Review how the lesson or unit was structured, including cognitive tension, exposure calibration, and revision opportunities.

3. Lesson X-Ray Analysis (20 minutes). Examine evidence across architectural layers such as purpose, tension, engagement, revision, and transfer.

4. Transfer and Judgment Review (15 minutes). Discuss whether students were asked to apply knowledge, defend reasoning, or adapt strategies under variation.

5. Energy Impact Discussion (10 minutes). Reflect on how the design affected teacher capacity and student cognitive engagement.

6. Redesign Commitments (15 minutes). Identify one structural adjustment to test in the next cycle.

7. Reflection and Next Steps (5 minutes). Summarize key insights and clarify what evidence will be examined in the following meeting.

When studio agendas remain centered on instructional architecture, collaborative time is more likely to support meaningful redesign rather than routine information exchange.

PLC Facilitation Scripts

Anchor collaborative conversations in instructional design. Professional learning communities often function most effectively when conversations remain centered on instructional architecture rather than task completion or logistical updates. Shared facilitation language can help PLC leaders guide discussions toward reasoning, transfer, and revision – the intellectual priorities emphasized throughout this book.

Scripts provide a consistent starting point that helps teams examine how learning experiences cultivate thinking. Over time, shared language can support coherence across PLC meetings and instructional teams.

Examples of facilitation prompts may include:

Opening Prompt: "Our focus today is not only whether students completed the task, but how the task supported adaptive reasoning."

Diagnostic Prompt: "Where did cognitive tension sustain student thinking, and where did it diminish?"

Iteration Prompt: "What structural adjustment might strengthen opportunities for transfer under ambiguity?"

AI Alignment Prompt: "How did AI tools function within this lesson? Did they reduce extraneous load, or did they unintentionally remove productive struggle?"

When PLC conversations consistently return to questions of design, teams are more likely to examine the architecture of learning rather than isolated classroom moments. Shared facilitation language can therefore help strengthen coherence across collaborative inquiry.

Leadership Reflection Logs

Make system learning visible over time. Sustained redesign efforts often benefit from leadership reflection that accompanies implementation cycles. When leaders document what is being tested, what evidence is emerging, and how system conditions are changing, the district's learning process becomes easier to track and refine. Reflection logs can therefore function as a simple structure that helps institutionalize inquiry rather than relying on informal memory.

Many districts use short, structured weekly reflections that focus on the architecture of learning rather than isolated events.

A leadership reflection log may include prompts such as:

Week's Design Focus: What architectural adjustment or redesign priority received attention during this cycle?

Evidence of Thinking: What observable indicators of student reasoning, transfer, or revision appeared in classrooms?

Energy Impact: Did the redesign influence teacher cognitive load or professional capacity? In what ways?

Equity Check: How was participation distributed across students? Did engagement patterns shift?

AI Integrity Check: Did AI tools support the intended learning architecture and remain aligned with the Ethical AI Decision Compass?

Documenting these reflections helps leadership teams observe patterns over time and maintain continuity across redesign cycles. When system learning is recorded and revisited, improvement efforts are more likely to sustain momentum rather than fading between initiatives.

AI Communication Templates

System redesign efforts often benefit from clear communication across the district community. When leaders explain the purpose of new initiatives and how AI integration supports learning goals, stakeholders are better able to understand the changes taking place. Transparent messaging can help reduce uncertainty while reinforcing the district's commitment to reasoning, ethical technology use, and sustainable teaching practices.

Communication may be tailored for different audiences so that each group understands how redesign efforts affect learning environments.

Board Communication Template: "Our district is implementing a 90-day redesign cycle focused on cultivating adaptive expertise in an AI-rich world. This initiative emphasizes cognitive depth, ethical AI integration, and sustainable teacher practice. Human judgment remains central to our instructional model."

Family Communication Template: "We are helping students learn how to use AI responsibly while continuing to develop independent reasoning. Some assignments will include AI disclosure expectations, and certain learning activities may involve device-free work so that students have opportunities for sustained thinking."

Teacher Communication Template: "This redesign cycle focuses on strengthening instructional architecture rather than increasing compliance expectations. AI tools may assist with reducing routine workload, while student reasoning and intellectual ownership remain central to learning design."

Clear communication helps stakeholders understand how technology integration aligns with educational goals. When expectations and purposes are visible, communities are more likely to engage with redesign efforts constructively and with confidence.

Implementation Principles

As districts move through redesign cycles, a small set of guiding principles can help maintain coherence across phases. These principles provide a consistent reference point for leadership teams as they test new structures, examine evidence, and refine system design. Rather than attempting broad transformation all at once, implementation often progresses through deliberate, cumulative adjustments.

Across redesign phases, districts may keep several priorities in view:

1. **Focus on one structural shift at a time.** Concentrated attention often allows teams to examine the effects of a change more clearly.
2. **Examine evidence of cognition rather than activity alone.** Observation and evaluation can attend to reasoning, explanation, and revision rather than only visible task completion.
3. **Protect teacher capacity.** Redesign efforts benefit from considering how system changes affect professional energy and workload.
4. **Integrate ethical AI guardrails early.** Clear expectations for responsible AI use help maintain alignment between technology integration and learning goals.
5. **Document learning across the system.** Recording insights and evidence helps ensure that redesign efforts contribute to collective improvement rather than remaining isolated experiments.

When these principles guide implementation, transformation tends to develop incrementally. Over time, small architectural adjustments accumulate into broader system change.

Leadership Reflection

The presence of artificial intelligence does not necessarily call for urgency alone. In many districts, the more pressing need is structure – processes that allow ideas to move from discussion into disciplined cycles of design, testing, and refinement.

Structured redesign cycles can help systems translate intention into practice. When districts examine evidence regularly, test adjustments in manageable pilots, and reflect on what is learned, improvement begins to accumulate over time. In this way, ideas continue to move forward, evidence builds, and professional energy can remain focused rather than scattered.

District transformation rarely emerges from declarations or isolated initiatives. More often, it develops through repeated cycles of design, observation, refinement, and expansion. Each cycle contributes small architectural adjustments that gradually strengthen coherence across the system.

In this context, the work of system leadership is less about chasing innovation and more about stewarding the architecture that supports learning. When redesign cycles remain visible and consistent, philosophy begins to translate into practice – and practice, in turn, becomes the foundation for sustainable improvement.

Superintendent Briefing: From Agreement to Action

Many districts recognize the need to redesign learning in an AI-rich environment. Agreement on this direction, however, does not automatically translate into change. System improvement typically occurs when leadership structures help convert shared ideas into consistent practices across schools and classrooms.

For this reason, some districts are beginning to organize improvement efforts through structured redesign cycles. Short implementation cycles allow leadership teams to examine areas of misalignment, test new approaches in practice, gather evidence, and refine strategies before expanding them more broadly.

The intention of a 90-day redesign cycle is not to resolve every challenge immediately. Instead, it provides a manageable structure that helps districts build momentum while learning from each iteration. Over time, repeated cycles allow insights to accumulate and adjustments to become more precise.

When districts adopt iterative redesign structures, instructional improvement often shifts from episodic initiatives toward continuous learning. In this way, change emerges not from inspiration alone, but from disciplined cycles of design, observation, and refinement that gradually strengthen the system.

System Leader Playbook: Implementing District Redesign

District redesign rarely emerges from isolated initiatives or one-time training sessions. More often, sustainable change develops when leadership structures help translate instructional philosophy into disciplined cycles of system improvement. The leadership actions below illustrate ways district teams can begin aligning structures, roles, and routines with the learning architecture described throughout this book.

1. **Begin with a Coherence Audit.**

Leadership teams may start by examining the five system layers introduced earlier: instructional design, professional expertise, ethical AI integration, system alignment, and leadership implementation.

This review can help identify where delivery-oriented structures still dominate and where system signals may conflict about the kinds of thinking schools intend to cultivate.

2. **Launch a 90-Day Redesign Cycle.**

Districts can organize improvement efforts around short redesign cycles that include four phases:

- Diagnose — examine areas of structural misalignment
- Pilot — test a focused redesign within a contained setting
- Iterate — refine structures using evidence from practice
- Scale — expand approaches that demonstrate promise

Short cycles often provide enough urgency to sustain momentum while preserving time for reflection and adjustment.

3. **Build cross-role leadership teams.**

Instructional redesign typically benefits from shared leadership. Teams may include instructional coaches, school administrators, teacher leaders, and technology or innovation specialists.

Cross-role collaboration helps ensure that redesign decisions reflect multiple perspectives and remain coherent across system layers.

4. **Anchor the system in a Graduate Profile.**

Districts may find clarity by articulating the capabilities they hope graduates will develop, such as: adaptive expertise, reasoning under ambiguity, ethical judgment, and transfer across domains.

When these outcomes are clearly defined, curriculum design, assessment practices, and technology integration can align more intentionally around them.

5. **Protect the conditions for deep work.**

Instructional redesign often requires time and professional focus. Leadership teams may therefore examine how schedules and expectations support:

- sustained planning time
- collaborative design work
- focused instructional attention

Without these conditions, system redesign can easily become superficial rather than structural.

Closing Reflection

District transformation rarely occurs through sudden breakthroughs. More often, it emerges through deliberate cycles that strengthen coherence across the system. When leadership actions remain aligned with shared intellectual goals – and when redesign proceeds through disciplined iteration – districts gradually build learning environments where reasoning, judgment, and sustained inquiry can flourish.

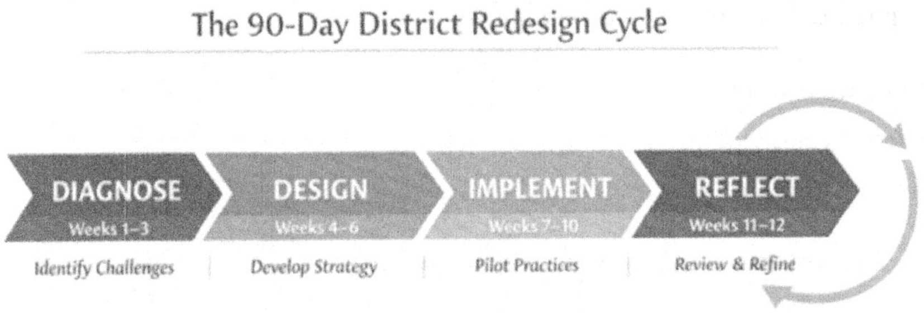

Figure 2: The 90-Day District Redesign Cycle. Four sequential stages designed to translate instructional philosophy into disciplined cycles of system diagnosis, experimentation, refinement, and scaled implementation.

The Designing Schools Framework

The Designing Schools framework organizes district redesign into five interconnected system layers. Each layer addresses a different dimension of how learning environments function. When these layers align, schools cultivate reasoning, adaptability, and judgment in an AI-rich world. When they operate independently, even strong initiatives fail to scale.

Including:

- Learning Architecture: How learning experiences are structured to cultivate reasoning rather than task completion.
- Adaptive Expertise: How educators develop the adaptive judgment required to design and refine learning environments.
- Ethical AI Integration: How artificial intelligence is governed so that it supports cognition and protects student dignity.
- System Alignment: How curriculum, assessment, grading, and accountability reinforce the same intellectual goals.
- Leadership Implementation: How districts build structures that sustain redesign over time.

District transformation does not occur through isolated initiatives. It occurs when instructional design, professional expertise, technology governance, system alignment, and leadership implementation operate as a coherent architecture.

This framework summarizes the system architecture described throughout the book. Each layer reinforces the others. When alignment occurs across all five layers, districts cultivate graduates capable of reasoning, adapting, and exercising judgment in an AI-rich world.

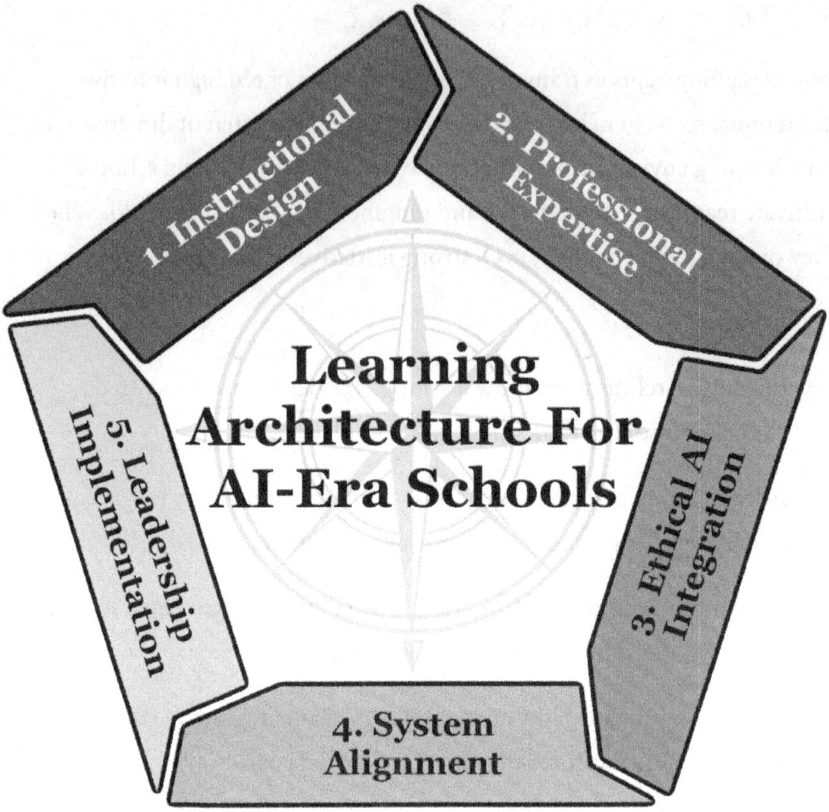

Figure 3: The Designing Schools Framework. Five interdependent layers of district redesign in an AI-rich learning system. Layers are (1) Instructional design, (2) professional expertise, (3) ethical AI integration, (4) system alignment, and (5) leadership implementation.

District Leader Action Summary

Designing Schools in an Age of Artificial Intelligence

Artificial intelligence has not changed the fundamental purpose of education. Schools still exist to cultivate thinking, judgment, and the capacity to learn continuously. What AI has changed is the environment in which those goals must be achieved.

When answers can be generated instantly, systems that prioritize coverage, completion, and polished output become fragile. The long-term advantage of schools lies not in producing information efficiently but in cultivating learners who can reason, interpret, and adapt when conditions change.

The Designing Schools framework presented in this book identifies five system layers that determine whether learning environments cultivate thinking or merely accelerate performance.

District redesign requires alignment across all five layers.

The Five Layers of School Redesign

System Layer	Core Leadership Question
Instructional Design	Are learning experiences structured to require explanation, reasoning, and transfer?
Professional Expertise	Are educators developing the adaptive expertise required to design learning environments?
Ethical AI Integration	Does technology support cognition while protecting student dignity?
System Alignment	Do curriculum, assessment, and grading reinforce the same intellectual goals?
Leadership Implementation	Are district structures sustaining continuous instructional redesign?

When these layers operate independently, improvement efforts remain fragmented. When they align, schools cultivate graduates capable of reasoning and exercising judgment in an AI-rich world.

Leadership Priorities for District Teams

District leaders do not need to redesign entire systems immediately. Improvement begins with focused architectural shifts that accumulate over time.

The following priorities represent high-leverage starting points for district leadership teams.

1. Redesign Tasks to Require Thinking

Examine assignments across the district. Ask:

- Do tasks require explanation and reasoning?
- Do they require transfer to unfamiliar contexts?
- Do they allow revision and defense of ideas?

When learning tasks reward thinking, students prioritize thinking.

2. Shift Assessment Toward Reasoning

Assessment systems strongly influence instructional practice. District leaders should ensure that assessments require:

- explanation of reasoning
- interpretation of evidence
- contextual judgment
- revision and refinement

Grades should reflect intellectual ownership rather than polished output alone.

3. Build Teacher Design Capacity

Instructional improvement depends on educators who can design learning environments intentionally. Districts should support teachers through:

- collaborative lesson design
- instructional coaching focused on architecture
- redesign studios within PLC structures
- iterative instructional experimentation

When design expertise becomes a shared professional skill, improvement becomes sustainable.

4. Govern Artificial Intelligence Responsibly

AI tools should support learning rather than replace student thinking. District leaders should ensure that:

- AI expectations are clearly communicated
- technology preserves student dignity
- educators retain responsibility for decisions
- tools are evaluated for bias and equity impact

Technology must serve educational values—not redefine them.

5. Use Iterative Improvement Cycles:

District transformation occurs through structured cycles of redesign. The 90-Day District Redesign Cycle introduced in Chapter 12 provides a manageable approach:

1. Diagnose instructional architecture
2. Design targeted improvements
3. Implement redesigned practices
4. Reflect on evidence and refine systems

Repeated cycles gradually strengthen instructional coherence across schools.

Signals of a Design-Oriented School System

District leaders will know redesign is occurring when classrooms increasingly demonstrate:

- students explaining their reasoning
- students revising work after feedback
- students applying ideas in unfamiliar contexts
- teachers discussing instructional architecture collaboratively
- learning environments where intellectual risk is safe

These signals reflect systems designed for thinking rather than performance alone.

Signals of a Delivery-Oriented System

In contrast, systems still operating primarily as delivery models often demonstrate:

- heavy emphasis on pacing and coverage
- assessments focused primarily on correct answers
- limited opportunities for revision
- classroom participation dominated by a few students
- assignments that AI tools can complete without reasoning

Recognizing these patterns allows leadership teams to identify redesign priorities.

The Work of System Leadership

School redesign in an AI-rich era does not require abandoning the principles of learning science that have guided education for decades. Instead, it requires aligning systems with those principles more intentionally.

- District leaders must ensure that:
- instructional design cultivates reasoning
- professional learning develops adaptive expertise
- technology governance protects student dignity
- assessment systems reinforce intellectual goals
- leadership structures sustain redesign over time

When these elements align, schools cultivate graduates capable of navigating a world where information is abundant but judgment remains scarce.

Final Leadership Question: Artificial intelligence can generate answers instantly. The enduring question for education therefore becomes:

Are our schools designed to produce answers – or to cultivate thinkers capable of reasoning when answers are abundant?

The systems that thrive in an AI-rich future will be those that design learning environments where human judgment remains central.

APPENDICES

Leadership Toolkit for Designing Schools

The following tools support district leaders, principals, and instructional teams in applying the Designing Schools framework to real instructional systems.

Each tool corresponds to one or more system layers described in the book and can be used independently or as part of a structured district redesign cycle.

District leadership teams are encouraged to begin with diagnostic tools that clarify current instructional architecture, then move toward design and implementation tools that support continuous improvement.

Appendix Overview

Appendix	Tool	Purpose
Appendix A	Instructional Coherence Audit	Diagnose whether instructional practices reflect shared design principles
Appendix B	Delivery-to-Design Diagnostic	Identify whether district systems prioritize information delivery or learning design
Appendix C	AI-Resilient Task Design Checklist	Evaluate whether learning tasks require student reasoning and ownership
Appendix D	Lesson Load Mapping Protocol	Analyze cognitive load and attention demands within lessons
Appendix E	90-Day District Redesign Planner	Guide district improvement through structured redesign cycles
Appendix F	Superintendent AI Governance Checklist	Evaluate AI tools for ethical and instructional alignment
Appendix G	Ethical AI Decision Compass	Support leadership teams in evaluating technology adoption decisions

How the Toolkit Aligns with the Designing Schools Framework

Each appendix supports one of the five system layers described in the book.

System Layer	Supporting Tools
Instructional Design	Appendices C and D
Professional Expertise	Appendices A and B
Ethical AI Integration	Appendices F and G
System Alignment	Appendices A and B

Together, these tools allow leadership teams to move from diagnosis to redesign while maintaining coherence across instructional, professional, and technological systems.

Suggested Implementation Sequence

District leaders may use the toolkit in the following sequence during improvement cycles.

Step 1 — Diagnose Instructional Architecture. Use diagnostic tools to understand current system patterns. Tools:

- Appendix A — Instructional Coherence Audit
- Appendix B — Delivery-to-Design Diagnostic

Step 2 — Redesign Learning Tasks and Lessons. Strengthen classroom learning architecture. Tools:

- Appendix C — AI-Resilient Task Design Checklist
- Appendix D — Lesson Load Mapping Protocol

Step 3 — Implement System Redesign Cycles. Pilot and evaluate instructional improvements. Tool:

- Appendix E — 90-Day District Redesign Planner

Step 4 — Evaluate Technology Decisions. Ensure AI tools align with district instructional values. Tools:

- Appendix F — Superintendent AI Governance Checklist
- Appendix G — Ethical AI Decision Compass

Leadership Guidance

These tools are not intended to create additional compliance requirements. They are designed to help leadership teams examine the structures that shape learning environments.

When used collaboratively, the toolkit can support:

- clearer instructional expectations across schools
- stronger alignment between curriculum and assessment
- responsible integration of artificial intelligence
- sustained cycles of instructional improvement

District redesign does not occur through isolated initiatives. It emerges when leaders consistently examine and refine the architecture of learning systems.

Final Leadership Reminder

The purpose of redesign is not to accelerate instruction.

It is to ensure that schools cultivate learners capable of reasoning under uncertainty, adapting knowledge across contexts, revising thinking through evidence, and exercising ethical judgment

The tools in this appendix support leadership teams in designing systems where those capabilities can develop.

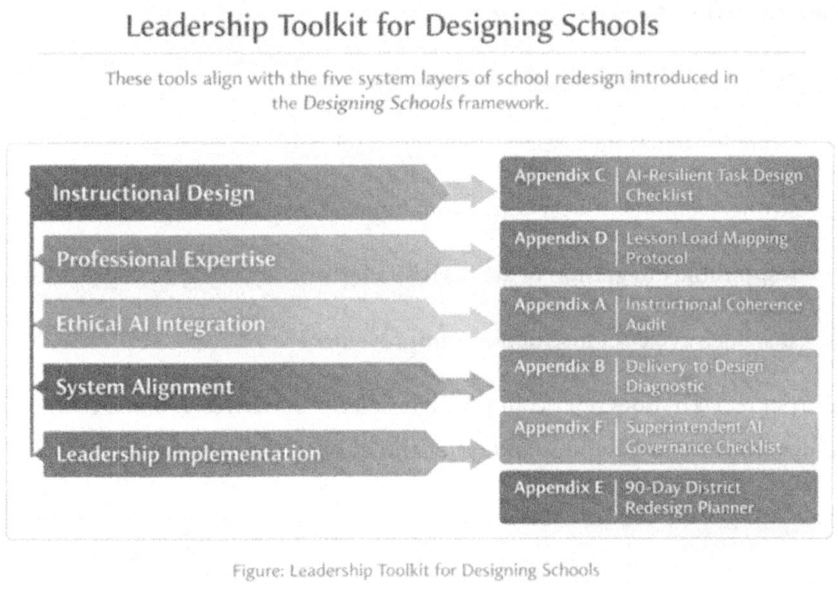

Figure 4: Leadership Toolkit for Designing Schools. A structured set of diagnostic and redesign tools aligned with the five system layers – Instructional Design, Professional Expertise, Ethical AI Integration, System Alignment, and Leadership Implementation – to help district teams translate learning science and ethical AI principles into coherent system practice.

Appendix A: Instructional Coherence Audit Tool

Purpose: The Instructional Coherence Audit helps district leadership teams determine whether classroom practices reflect a shared instructional design philosophy or a collection of isolated practices. In AI-rich learning environments, coherence becomes essential. When instructional architecture varies widely across classrooms, students encounter inconsistent expectations for thinking, reasoning, and revision.

This audit helps leadership teams examine whether the district's instructional systems consistently cultivate conceptual understanding, adaptive reasoning, and intellectual ownership rather than task completion and procedural compliance.

The goal is not uniformity of instruction. The goal is alignment around shared design commitments that promote deep learning.

How to Use the Audit

This tool was designed to be used by district leadership teams, school leadership teams, instructional coaches, professional learning communities, etc.

The audit should be conducted using multiple sources of evidence, including:

- Classroom observations
- Curriculum materials
- Student work samples
- Assessment design
- Teacher planning documents
- PLC discussions

Rate each domain using the following scale:

Rating	Description
1	Not Evident – Practice rarely observed across classrooms
2	Emerging – Practice appears inconsistently
3	Developing – Practice appears regularly but unevenly
4	Established – Practice appears consistently across classrooms
5	Embedded – Practice is deeply integrated into district culture

Leadership teams should discuss evidence for each rating and record supporting observations.

Domain 1: Cognitive Architecture

Examines whether learning experiences require students to think, reason, and construct understanding.

Indicator	Evidence Questions	Rating
Conceptual Understanding	Do lessons require students to explain underlying concepts rather than only produce answers?	
Cognitive Tension	Do lessons include problems, questions, or contradictions that require reasoning?	
Productive Struggle	Are students given time to wrestle with ideas before receiving explanation?	
Visible Thinking	Do students regularly articulate their reasoning verbally or in writing?	
Transfer Opportunities	Are students asked to apply knowledge in unfamiliar contexts?	

Leadership Reflection: Are classrooms designed for thinking or task completion? Do students regularly experience intellectual challenge?

Domain 2: Assessment Alignment

Examines whether assessments measure reasoning and understanding rather than polished output alone.

Indicator	Evidence Questions	Rating
Explanation Requirements	Do assessments require students to explain reasoning?	
Transfer Tasks	Do assessments require application in new contexts?	
Defense of Thinking	Are students asked to justify decisions or interpretations?	
Revision Opportunities	Are students allowed to revise work after feedback?	
AI-Resistant Design	Would students still need to reason if AI tools were removed?	

Leadership Reflection: Do grades reflect thinking or task completion? What kind of learning does our assessment system reward?

Domain 3: Instructional Design Consistency

Examines whether classrooms share common design principles across grade levels and schools.

Indicator	Evidence Questions	Rating
Shared Design Language	Do educators use common terms (e.g., cognitive tension, transfer, revision)?	
Unit Architecture	Are learning experiences intentionally sequenced around reasoning and inquiry?	
Vertical Alignment	Does cognitive demand increase across grade levels?	
Structured Inquiry	Are students expected to ask questions and investigate ideas?	
Revision Culture	Is iteration a normal part of learning across classrooms?	

Leadership Reflection: Do students encounter consistent expectations for thinking across classrooms? Is instructional design coherent across the district?

Domain 4: Classroom Culture for Thinking

Examines whether classroom environments make cognitive risk safe and sustainable.

Indicator	Evidence Questions	Rating
Psychological Safety	Do students feel safe sharing uncertain thinking?	
Error as Data	Are mistakes treated as opportunities for learning?	
Participation Equity	Are multiple students contributing to discussion and reasoning?	
Exposure Calibration	Are students given rehearsal time before public responses?	
Belonging Signals	Do classrooms communicate that intellectual effort is valued?	

Leadership Reflection: Do students feel safe enough to attempt difficult thinking? Are participation structures equitable?

Domain 5: AI and Cognitive Integrity

Examines whether AI integration supports cognition rather than replacing it.

Indicator	Evidence Questions	Rating
AI Transparency	Are expectations for AI use clearly communicated to students?	
Thinking Preservation	Do assignments require reasoning that cannot be fully outsourced?	
Productive AI Use	Is AI used to reduce extraneous load rather than replace thinking?	
Intellectual Ownership	Can students explain work they submit?	
Ethical Integration	Do AI policies protect student dignity and learning integrity?	

Leadership Reflection: Does AI strengthen or weaken student thinking in our classrooms?

Identifying System Patterns

After completing the audit, leadership teams should examine patterns across domains. Questions for reflection:

- Which domains show the strongest alignment?
- Where is instructional design inconsistent?
- Are expectations for thinking clear across classrooms?
- Which structures reinforce compliance rather than reasoning?

Warning Indicators of Low Instructional Coherence

Districts with weak instructional coherence can demonstrate the following patterns: High variability in cognitive demand across classrooms. Heavy emphasis on pacing and coverage. Assessments focused primarily on correct answers. Limited student explanation or defense of reasoning. Minimal opportunities for revision. Classroom participation dominated by a small number of students

When these patterns appear, instructional architecture may depend on individual teacher practice rather than system design.

Next Steps for Leadership Teams

After completing the audit, leadership teams should identify one structural improvement priority for the next instructional cycle. Possible next steps include:

- Redesigning assessments to require explanation and transfer
- Revising observation tools to measure cognitive architecture
- Embedding revision opportunities into curriculum
- Aligning PLC conversations around instructional design
- Clarifying expectations for AI-supported learning

Instructional coherence improves when systems focus on one architectural shift at a time, observe results, and refine practice through iterative cycles. Key Leadership Question: Are our instructional systems designed to produce students who complete tasks – or students who can reason when conditions change?

Appendix B: Delivery-to-Design Diagnostic

Purpose: For more than a century, most school systems have operated under a delivery model of instruction. In this model, curriculum is distributed, teachers deliver content, and student performance is measured through completed assignments and assessments.

Artificial intelligence exposes the limitations of this structure. When systems focus primarily on content delivery and polished output, AI tools can replicate many of those outcomes instantly.

Design-oriented systems operate differently. Rather than organizing instruction around the delivery of information, they engineer learning environments that cultivate reasoning, judgment, and conceptual understanding.

The Delivery-to-Design Diagnostic helps district leadership teams examine where their systems currently fall along this continuum.

How to Use the Diagnostic

This tool was designed to be used by Superintendent cabinet teams, district leadership teams, school leadership teams, instructional leadership teams, etc. Teams should review each domain and determine where current practice most closely aligns.

Each domain should be rated using the following scale:

Rating	Description
1	Strongly Delivery-Oriented
2	Mostly Delivery-Oriented
3	Transitional
4	Mostly Design-Oriented
5	Strongly Design-Oriented

Teams should record evidence supporting their rating and identify potential redesign priorities.

Domain 1: Curriculum Design

Delivery Model Signal	Design Model Signal	Rating Evidence
Curriculum organized around topic coverage	Curriculum organized around conceptual understanding	
Units emphasize completion of assignments	Units emphasize explanation and reasoning	
Learning goals focus on correct answers	Learning goals focus on interpretation and transfer	
Activities prioritize efficiency and coverage	Activities prioritize inquiry and exploration	

Leadership Reflection: Does our curriculum prioritize exposure to information or construction of understanding?

Domain 2: Assessment Practices

Delivery Model Signal	Design Model Signal	Rating Evidence
Assessments measure correct answers	Assessments measure reasoning and explanation	
Polished final products determine grades	Revision and refinement are part of learning	
Tasks can be completed independently with AI tools	Tasks require contextual reasoning and defense	
Performance is measured primarily through written output	Performance includes dialogue, defense, and application	

Leadership Reflection: Do our assessments reward thinking or production of answers?

Domain 3: Instructional Practice

Delivery Model Signal	Design Model Signal	Rating Evidence
Teacher explanation dominates class time	Student reasoning is visible and frequent	
Students practice procedures repeatedly	Students apply knowledge under variation	
Lessons emphasize speed and completion	Lessons emphasize explanation and reflection	
Cognitive challenge depends on individual teacher expertise	Cognitive challenge is intentionally designed into curriculum	

Leadership Reflection: Are students primarily receiving information or constructing meaning?

Domain 4: Classroom Culture

Delivery Model Signal	Design Model Signal	Rating Evidence
Correct answers are prioritized	Reasoning processes are valued	
Errors are corrected quickly	Errors are examined and discussed	
Participation is dominated by confident students	Participation structures distribute thinking widely	
Students protect identity by avoiding mistakes	Students engage in intellectual risk-taking	

Leadership Reflection: Do classrooms protect performance or cultivate intellectual growth?

Domain 5: Professional Learning

Delivery Model Signal	Design Model Signal	Rating Evidence
Professional learning focuses on strategies	Professional learning focuses on instructional design	
Teachers implement programs	Teachers analyze learning architecture	
PLC meetings review pacing and scores	PLC meetings analyze cognitive demand and transfer	
Improvement depends on individual teacher innovation	Improvement emerges through collaborative design	

Leadership Reflection: Are educators treated primarily as implementers or designers?

Domain 6: System Leadership

Delivery Model Signal	Design Model Signal	Rating Evidence
Leadership focuses on compliance and pacing	Leadership focuses on learning architecture	
Observation systems emphasize engagement and management	Observation systems examine reasoning and cognitive demand	
Accountability focuses on coverage and scores	Accountability examines thinking and transfer	
Innovation occurs through isolated initiatives	Redesign occurs through coordinated system architecture	

Leadership Reflection: Does leadership reinforce delivery or design?

Interpreting Results

After completing the diagnostic, calculate an overall pattern:

Average Score	Interpretation
1-2	System operates primarily as a delivery model
3	System is transitioning toward design orientation
4-5	System consistently prioritizes learning design

Most districts will fall in the transitional range, where elements of both models coexist. This is normal. The goal of the diagnostic is not perfection but clarity about redesign priorities.

Identifying Redesign Priorities

Leadership teams should discuss the following questions:

- Which domains show the strongest delivery patterns?
- Which domains already demonstrate design-oriented practices?
- Where could one structural redesign produce the greatest impact?

District improvement rarely occurs through sweeping reform. It occurs through focused architectural shifts that accumulate over time.

Examples include:

- Converting assessments to require explanation and defense
- Redesigning units to include transfer tasks
- Shifting PLC meetings toward lesson architecture analysis
- Revising observation tools to examine cognitive demand

Leadership Action Prompt

Select one domain as the focus of the next instructional redesign cycle.

Ask:

- What structural change would move this domain one step toward design?
- What evidence would show improvement?
- How will we observe the change across classrooms?

Design-oriented systems do not emerge through slogans or initiatives. They emerge through deliberate redesign of instructional architecture.

Leadership Question

Is our system organized to deliver information efficiently – or to cultivate thinkers capable of reasoning when answers are abundant?

Appendix C: AI-Resilient Task Design Checklist

Purpose: Artificial intelligence can now generate essays, solve equations, summarize texts, and produce explanations instantly. When assignments rely primarily on information retrieval or polished output, AI tools can often complete them with minimal student reasoning.

This does not make assignments meaningless—but it changes what educators must design for.

AI-resilient learning tasks require students to demonstrate ownership of thinking. They require explanation, contextual judgment, adaptation, and revision.

This checklist helps educators examine whether a learning task cultivates human reasoning rather than simply producing answers.

The goal is not to eliminate AI tools from classrooms. The goal is to ensure that learning tasks still require student cognition.

How to Use This Checklist

This tool was designed to be used alongside educators. Before implementing a lesson or assessment, educators should review the task using the following questions.

For each item, mark:

✓ Yes – The task requires this element
△ Partially – The element is present but limited
✕ No – The element is absent

Tasks that receive mostly Yes responses are generally resilient to automation and promote deeper learning.

Section 1: Reasoning and Explanation

Does the task require students to demonstrate thinking rather than simply produce answers?

Question	✓	△	✕
Students must explain how they reached their conclusions			
Students must justify reasoning using evidence or logic			
Students must compare alternative interpretations or strategies			
Students must articulate the conceptual idea behind the task			

Design Principle: Tasks should reward explanation and reasoning, not just correct outcomes.

Section 2: Contextual Judgment

Does the task require students to make decisions within real or simulated contexts?

Question	✓	△	✕
The task includes constraints or competing priorities			
Students must weigh different possible solutions			
Students must defend a recommendation or decision			
Students must interpret ambiguous information			

Design Principle: AI can produce answers. Judgment emerges when context complicates those answers.

Section 3: Transfer and Adaptation

Does the task require students to apply knowledge beyond familiar examples?

Question	✓	△	✕
Students apply ideas in a new context			
Students must adapt strategies when conditions change			
Students encounter unfamiliar problem conditions			
Students must connect ideas across disciplines			

Design Principle: Understanding is demonstrated through transfer, not repetition.

Section 4: Ownership of Thinking

Can students demonstrate that the thinking is genuinely theirs?

Question	✓	△	✕
Students must explain reasoning verbally or in discussion			
Students must respond to questions about their work			
Students must revise thinking after critique			
Students must defend ideas publicly or in writing			

Design Principle: Ownership emerges when students must articulate and defend their thinking.

Section 5: Revision and Reflection

Does the task include opportunities for improvement and metacognition?

Question	✓	△	✕
Students receive feedback before final submission			
Students revise work based on critique			
Students reflect on how their thinking changed			
Students analyze mistakes or alternative approaches			

Design Principle: Learning deepens when students revise ideas rather than submit final answers immediately.

Section 6: Cognitive Integrity with AI

Does the task ensure that AI supports thinking rather than replacing it?

Question	✓	△	✕
AI use expectations are clearly defined			
Students must demonstrate reasoning even if AI tools are used			
AI reduces peripheral workload rather than core thinking			
Students remain responsible for interpreting results			

Design Principle: AI should reduce extraneous effort, not eliminate germane cognitive work.

Quick AI-Resilience Test

Ask the following question: If a student used AI to complete this task, would meaningful thinking still be required?

If the answer is no, redesign the task to include

- explanation
- contextual judgment
- transfer
- defense
- revision

Example Redesign

Low-Resilience Task: Write a five-paragraph essay explaining the causes of the American Revolution. AI can easily generate this response.

AI-Resilient Redesign: Advise a colonial assembly deciding whether to support independence.
Present your recommendation and defend it against counterarguments from opposing delegates.

Students must:

- interpret historical context
- weigh competing values
- defend reasoning
- respond to critique

Design Reminder

Strong learning tasks do not eliminate artificial intelligence.
They redefine what counts as understanding.

When tasks require explanation, transfer, revision, and judgment, AI becomes a tool for learning rather than a substitute for thinking.

Design Question for Educators: Does this task reward answer production – or the development of human judgment?

Appendix D: Lesson Load Mapping Protocol

Purpose: Learning depends on how instructional design manages cognitive load. Human working memory is limited, and when learning tasks overwhelm that capacity, students struggle to construct understanding.

Cognitive Load Theory distinguishes three types of cognitive load that influence learning:

Intrinsic Load: The inherent complexity of the material being learned.

Extraneous Load: Cognitive burden created by poor instructional design.

Germane Load: The productive mental effort devoted to constructing understanding.

Strong instructional design reduces unnecessary load while preserving the effort required for meaningful thinking. In AI-rich classrooms, this balance becomes even more important. Artificial intelligence can reduce unnecessary workload, but it can also eliminate the very cognitive effort that strengthens learning.

The Lesson Load Mapping Protocol helps educators analyze instructional tasks before implementation so they can intentionally design for clarity, productive struggle, and conceptual understanding.

How to Use the Protocol

This tool was designed to be used by:

- Individual teachers during lesson planning
- Instructional coaches during planning conferences
- Professional learning communities during lesson study
- District curriculum teams reviewing instructional materials

For an upcoming lesson, analyze each component using the three categories of cognitive load.

Step 1: Identify Intrinsic Load

Intrinsic load reflects the natural complexity of the content. Some concepts are inherently more demanding than others and require more cognitive effort to understand.

Reflection Questions	Notes
What core concept must students understand in this lesson?	
Is the concept abstract, layered, or unfamiliar?	
What prior knowledge do students need to engage successfully?	
Which parts of the content are most cognitively demanding?	

Planning Considerations

To manage intrinsic load:

- Break complex ideas into manageable steps
- Connect new ideas to prior knowledge
- Provide examples that illustrate the concept clearly
- Sequence learning tasks intentionally

The goal is not to remove complexity, but to introduce complexity in a manageable progression.

Step 2: Identify Extraneous Load

Extraneous load is created by instructional design that distracts from learning.

Examples include unclear directions, unnecessary steps, confusing visuals, or constant task switching.

Reflection Questions	Notes
Are instructions concise and clearly sequenced?	
Do visuals and materials support the learning goal?	
Are there unnecessary steps in the task?	
Are students required to multitask across too many elements?	
Does technology introduce distractions or confusion?	

Common Sources of Extraneous Load

- overly complex instructions
- excessive worksheets or steps
- visual clutter on slides or materials
- constant task switching
- unclear expectations
- unmanaged device use

Design Goal: Remove obstacles that consume attention without contributing to understanding.

Step 3: Identify Germane Load

Germane load represents the productive thinking that builds understanding.

These are the moments where students organize ideas, explain reasoning, and make connections.

Reflection Questions	Notes
Where must students explain their thinking?	
Where must students retrieve prior knowledge?	
Where must students apply ideas in a new situation?	
Where must students revise or refine their thinking?	
Where must students defend reasoning or interpret ideas?	

Examples of Germane Load

- explaining a concept in one's own words
- comparing alternative strategies
- analyzing mistakes
- applying knowledge in unfamiliar contexts
- revising explanations after feedback

Design Goal: Protect and expand opportunities for productive intellectual effort.

Step 4: Evaluate AI's Role in the Lesson

AI tools can influence all three forms of cognitive load. Educators must examine whether AI supports thinking or replaces it.

Reflection Questions	Notes
Does AI reduce unnecessary workload (extraneous load)?	
Does AI remove the struggle required for understanding (germane load)?	
Are students still required to explain reasoning?	
Do students remain responsible for interpreting results?	

Healthy AI Use

AI can support learning when it:

- summarizes complex texts before analysis
- provides practice examples
- clarifies vocabulary
- generates feedback prompts

AI undermines learning when it:

- generates full responses students do not understand
- replaces the first attempt at reasoning
- removes the need for explanation or revision

Design goal: Use AI to reduce unnecessary cognitive burden while preserving the thinking required for learning. AI should support schema development, not substitute for student reasoning.

Step 5: Map the Lesson Architecture

Summarize how the lesson balances the three forms of cognitive load.

Component	Evidence in Lesson
Intrinsic Load (conceptual complexity)	
Extraneous Load removed	
Germane Load preserved	
AI used productively	

Redesign If Necessary

If analysis shows the lesson:

- overwhelms students with extraneous complexity
- removes productive struggle
- relies on polished output rather than reasoning

then redesign the lesson to strengthen cognitive clarity and thinking ownership.

Possible adjustments include:

- simplifying instructions
- reducing task fragmentation
- inserting explanation checkpoints
- requiring reasoning before answers
- adding revision opportunities

Example Lesson Load Map

Lesson Topic: Ecosystem Interdependence

Load Type	Lesson Design
Intrinsic Load	Students analyze how changes to predator populations affect ecosystem balance
Extraneous Load	Simplified diagrams and step-by-step modeling reduce confusion
Germane Load	Students predict outcomes, defend reasoning, and revise explanations after new evidence
AI Role	AI used to generate additional ecosystem scenarios for analysis

Design goal: Ensure the lesson balances clarity and challenge so that students can focus attention on reasoning, explanation, and conceptual understanding rather than navigating unnecessary complexity.

Design Reminder

Effective instructional design does not eliminate difficulty.
It ensures that difficulty occurs in the right place.

Remove the obstacles that waste attention.
Preserve the effort that builds understanding.

Key Leadership Question: Does this lesson reduce unnecessary cognitive burden while preserving the thinking that leads to learning?

Appendix E: Adaptive Expertise Observation Rubric

The 90-Day District Redesign Planner

Purpose: Sustainable instructional improvement rarely occurs through isolated initiatives or rapid reforms. Systems improve when leaders focus on specific architectural shifts, observe evidence of change, and refine practice through iterative cycles.

The 90-Day District Redesign Planner helps leadership teams translate the Designing Schools framework into manageable improvement cycles. Rather than attempting to redesign entire systems at once, districts focus on one structural improvement, implement changes deliberately, and examine evidence before scaling further.

Each cycle includes four phases:

1. Diagnose
2. Design
3. Implement
4. Reflect and Refine

The cycle emphasizes learning through iteration rather than immediate perfection.

Overview of the 90-Day Cycle

Phase	Focus	Timeframe
Diagnose	Identify instructional architecture challenges	Weeks 1–3
Design	Develop targeted redesign strategies	Weeks 4–6
Implement	Pilot redesigned practices	Weeks 7–10
Reflect & Refine	Analyze evidence and adjust systems	Weeks 11–12

Each phase builds toward the next, allowing leadership teams to observe how changes influence teaching and learning.

Phase 1: Diagnose (Weeks 1–3)

The diagnostic phase focuses on identifying where instructional architecture requires redesign.

Leadership teams should examine multiple forms of evidence, including:

- classroom observations
- student work samples
- curriculum materials
- assessment tasks
- teacher feedback
- student engagement patterns

Diagnostic Questions

Questions	Evidence
Where do students experience the greatest difficulty reasoning or transferring knowledge?	
Which assignments can be completed with minimal student reasoning?	
Where do assessments measure correctness but not explanation?	
Where do classrooms rely on teacher explanation rather than student reasoning?	
Where does cognitive demand vary widely across classrooms?	

Leadership teams are encouraged to use tools such as:

- Instructional Coherence Audit (Appendix A)
- Delivery-to-Design Diagnostic (Appendix B)

Outcome of Phase 1: Identify one instructional architecture priority for the redesign cycle.

Examples may include:

- increasing opportunities for student explanation
- redesigning assessments for transfer
- strengthening revision culture
- clarifying expectations for AI-supported learning

Focusing on one priority allows the system to concentrate effort and observe results clearly.

Phase 2: Design (Weeks 4–6)

Once a priority is identified, leadership teams collaborate with instructional leaders and educators to design targeted improvements.

Design Questions

Questions	Evidence
What structural change would address the identified challenge?	
Which classrooms or teams will pilot the redesign?	
What instructional supports will educators need?	
How will success be measured?	

Possible redesign strategies include:

- embedding explanation checkpoints into lessons
- converting assessments into transfer tasks
- introducing structured revision cycles
- redesigning assignments to require defense of reasoning
- adjusting device policies to protect focused thinking time

Leadership teams should also determine:

- what evidence will signal improvement
- what supports teachers require during implementation

Outcome of Phase 2: A clear redesign plan for implementation across selected classrooms or teams.

Phase 3: Implement (Weeks 7–10)

During this phase, educators pilot the redesigned instructional structures.

Leadership teams should monitor implementation using:

- classroom walkthroughs
- teacher reflection meetings
- student work analysis
- coaching conversations

Evidence Collection

Evidence Source	Observations
Student explanations and reasoning	
Evidence of revision and feedback	
Student participation in discussion	
Transfer of learning across contexts	
Teacher feedback on instructional design	

The purpose of this phase is not to judge success immediately but to observe how redesigned structures influence learning behavior.

Leadership teams should ask:

- Are students reasoning more frequently?
- Are classrooms demonstrating greater cognitive engagement?
- Are teachers experiencing manageable implementation demands?

Phase 4: Reflect and Refine (Weeks 11–12)

The final phase focuses on analyzing evidence from the implementation phase and refining system practices.

Leadership teams should review:

- classroom observation data
- student work samples
- teacher feedback
- assessment results

Reflection Questions

Questions	Evidence
Where did the redesign strengthen student thinking?	
Where did implementation challenges appear?	
What structural adjustments would improve results?	
Should the redesign expand to additional classrooms or schools?	

The purpose of reflection is not to evaluate individuals but to improve instructional architecture.

Outcome of Phase 4: Leadership teams decide whether to:

- scale the redesign across the district
- refine the strategy and test again
- shift focus to a new priority

Cycle Planning Template

Leadership teams may use the following template to document each redesign cycle.

Component	Notes
Instructional Priority	
Diagnostic Evidence	
Redesign Strategy	
Pilot Classrooms or Schools	
Implementation Timeline	
Evidence of Impact	
Next Steps	

Leadership Guidance

District transformation occurs through structured iteration rather than sweeping reform. When leaders focus on one architectural improvement at a time, systems develop stronger instructional coherence and professional expertise.

The 90-Day District Redesign Cycle ensures that improvement efforts remain focused, evidence-based, manageable for educators, and aligned with district learning goals.

Over time, repeated cycles build sustainable instructional capacity across the system.

Leadership Reflection; Are our improvement efforts driven by new initiatives – or by deliberate redesign of the structures that shape learning?

Appendix F: Superintendent AI Governance Checklist

Purpose: Artificial intelligence is entering school systems rapidly through instructional tools, grading systems, predictive analytics platforms, and administrative technologies. While these tools can increase efficiency and expand access to information, they also introduce significant ethical and instructional considerations.

District leaders must ensure that artificial intelligence supports the educational mission of schools rather than quietly reshaping it. AI systems should strengthen learning, protect student dignity, and preserve human judgment in educational decision-making.

This checklist helps superintendents and district leadership teams evaluate whether AI tools and policies align with the core principles of responsible educational leadership.

The checklist is not designed to prevent innovation. It is designed to ensure that innovation occurs with clarity, transparency, and ethical responsibility.

How to Use This Checklist

This was designed for district leadership teams to use when:

- evaluating new AI tools or platforms
- developing district AI policies
- reviewing technology procurement decisions
- assessing classroom AI integration practices
- communicating expectations to staff and families

For each question, mark:

✓ Yes – Clear evidence of alignment
△ Partial – Some alignment but further work needed
✗ No – Significant concern requiring attention

Section 1: Instructional Integrity

Does the use of AI preserve meaningful student thinking and learning?

Question	✓	△	✕
Does the tool support learning rather than replace student reasoning?			
Do assignments still require explanation and conceptual understanding?			
Are students responsible for interpreting and defending their work?			
Does AI reduce extraneous workload rather than eliminate productive struggle?			

Leadership Reflection: AI should support cognition. It should not replace the intellectual work required for learning.

Section 2: Student Dignity and Psychological Safety

Does the technology respect student identity and protect psychological safety?

Question	✓	△	✕
Does the tool avoid excessive monitoring or surveillance?			
Are students protected from public algorithmic labeling or ranking?			
Does the technology support participation rather than increase anxiety?			
Are students treated as learners rather than data points?			

Leadership Reflection: Technology should strengthen relationships within schools—not replace them.

Section 3: Transparency and Communication

Are students, educators, and families informed about how AI is used?

Question	✓	△	✕
Are expectations for AI use clearly communicated to students and staff?			
Are families informed about data collection and AI-supported tools?			
Is the role of AI in grading or decision-making transparent?			
Are human educators responsible for final decisions?			

Leadership Reflection: Trust in educational systems requires clear communication about technology use.

Section 4: Equity and Bias Considerations

Does the AI system protect equity and avoid amplifying existing disparities?

Question	✓	△	✕
Has the district examined potential bias in the tool's outputs?			
Does the tool function effectively across diverse linguistic and cultural contexts?			
Are students with disabilities considered in design and accessibility?			
Does the tool expand opportunity rather than reinforce inequality?			

Leadership Reflection: AI systems inherit patterns from the data used to train them.
Leaders must ensure these patterns do not reinforce inequity.

Section 5: Human Judgment and Accountability

Does the district maintain human responsibility for educational decisions?

Question	✓	△	✕
Are educators responsible for interpreting AI-generated insights?			
Does AI inform decisions rather than make them autonomously?			
Are educators trained to evaluate AI outputs critically?			
Does the district maintain clear accountability for outcomes?			

Leadership Reflection: Artificial intelligence can support decision-making. It must never replace professional judgment.

Section 6: Implementation Readiness

Is the district prepared to implement AI tools responsibly?

Question	✓	△	✕
Have educators received guidance on appropriate AI use?			
Does the district have clear policies governing AI use in instruction?			
Are data privacy protections clearly defined?			
Is there a process for reviewing new AI tools before adoption?			

Leadership Reflection: Technology adoption should follow intentional policy and leadership review, not convenience.

AI Governance Decision Prompt

Before adopting or expanding any AI tool, leadership teams should ask:

1. Does this technology strengthen student thinking?
2. Does it preserve dignity and trust?
3. Is its operation transparent to educators, students, and families?
4. Does human judgment remain central to decision-making?

If these conditions are not met, further evaluation is necessary before adoption.

Governance Principle

Artificial intelligence should expand the capacity of schools to cultivate thinking, judgment, and ethical reasoning.

It should never reduce education to automated efficiency.

Strong districts do not simply adopt technology.
They govern it intentionally.

Leadership Question: Does this technology strengthen the human work of education – or quietly replace it?

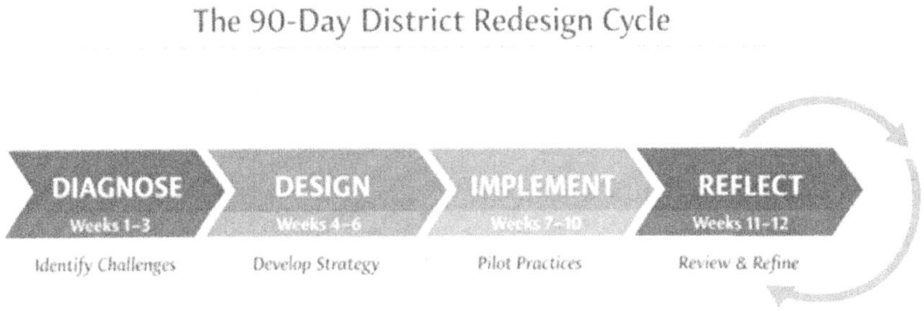

Figure 5: The 90-Day District Redesign Cycle. Four sequential stages designed to translate instructional philosophy into disciplined cycles of system diagnosis, experimentation, refinement, and scaled implementation.

Appendix G: Ethical AI Decision Compass

Purpose: Artificial intelligence tools are entering schools rapidly through writing assistants, adaptive learning systems, grading platforms, and predictive analytics dashboards. While these tools offer new possibilities for efficiency and personalization, they also introduce important instructional, ethical, and governance considerations.

District leaders must ensure that artificial intelligence strengthens learning environments rather than quietly reshaping them in ways that undermine student thinking, professional judgment, or educational equity.

The Ethical AI Decision Compass provides a structured framework for evaluating whether proposed AI tools align with the core purposes of education. Rather than focusing only on technical capability or efficiency, this tool helps leaders examine how AI influences cognition, dignity, equity, and human responsibility within learning systems.

The goal is not to prevent innovation. The goal is to ensure that innovation occurs within clear educational values.

How to Use the Decision Compass

This framework is designed for district technology committees, superintendent cabinet teams, curriculum leadership teams, instructional technology departments, and school leadership teams evaluating new tools

Before adopting or expanding an AI tool, leaders should examine the proposal using the four compass questions below. Each question highlights a critical dimension of responsible AI integration.

For each domain, leadership teams should ask:

- What benefits does the technology provide?
- What risks might emerge?
- What safeguards should be implemented?

Compass Question 1: Does This Tool Reduce or Replace Thinking?

AI can reduce unnecessary workload, but it can also remove the cognitive effort required for learning.

Some technologies reduce extraneous load, allowing students to focus on reasoning. Others eliminate germane cognitive work, replacing the thinking that builds understanding.

Indicators of Responsible Use

AI use is appropriate when it:

- simplifies complex texts before analysis
- supports brainstorming and idea generation
- provides additional practice opportunities
- accelerates feedback that encourages revision
- assists with language translation or accessibility

AI use becomes problematic when it:

- generates complete responses students cannot explain
- eliminates the need for first attempts at reasoning
- replaces the drafting and revision process
- produces polished work without conceptual understanding

Leadership Question: If the AI system were removed, would students still be able to perform the intellectual work required by the task?

If the answer is no, redesign may be necessary.

Compass Question 2: Does the Tool Widen or Narrow Identity Risk?

Identity risk refers to the vulnerability students experience when thinking publicly in learning environments. AI tools can either reduce or intensify this vulnerability.

Some technologies support participation by allowing students to rehearse ideas privately before sharing. Others increase surveillance or public comparison, raising anxiety and discouraging intellectual risk-taking.

Indicators of Responsible Use

AI supports identity safety when it:

- allows private rehearsal of thinking
- provides formative feedback without public ranking
- helps students clarify ideas before discussion
- scaffolds participation for hesitant learners

AI increases identity risk when it:

- publicly flags students as struggling or suspicious
- ranks students algorithmically
- monitors student behavior excessively
- replaces relational trust with automated detection

Leadership Question: Does this technology protect student dignity and psychological safety, or does it increase pressure and surveillance?

Compass Question 3: Does the Tool Privilege Speed Over Depth?

Artificial intelligence excels at generating responses quickly. While speed can increase efficiency, learning often requires slow thinking, revision, and reflection.

Systems that prioritize rapid output risk weakening the processes that develop understanding.

Indicators of Responsible Use

AI supports depth when it:

- helps students analyze complex information
- generates questions for deeper inquiry
- supports revision cycles
- encourages reflection on ideas

AI may undermine depth when it:

- encourages rapid answer production
- bypasses reading or analysis
- reduces opportunities for revision
- emphasizes efficiency over exploration

Leadership Question: Does this technology help students think more deeply, or simply produce answers faster?

Compass Question 4: Does the Tool Strengthen or Weaken Human Responsibility?

Education depends on human judgment. Teachers interpret student thinking, respond to emotional cues, and make ethical decisions about instruction and assessment.

AI systems must support — not replace — this professional responsibility.

Indicators of Responsible Use

AI strengthens professional practice when it:

- provides insights that educators interpret
- supports teacher planning and feedback
- assists with administrative workload
- enhances human decision-making

AI weakens professional responsibility when it:

- automates grading without teacher review
- generates instructional decisions without educator input
- replaces teacher feedback entirely
- obscures how conclusions are generated

Leadership Question: Are educators still responsible for interpreting and acting on the information provided by this system?

Decision Compass Summary

District leaders should examine any AI tool across the following four dimensions.

Compass Direction	Core Question
Cognition	Does this tool reduce or replace thinking?
Dignity	Does it protect or increase identity risk?
Depth	Does it support reflection or privilege speed?
Responsibility	Does it strengthen or weaken human judgment?

If concerns emerge in multiple domains, leaders should reconsider adoption or establish safeguards before implementation.

Implementation Safeguards

When districts choose to adopt AI tools, the following safeguards help protect learning integrity.

1. Clear Usage Expectations: Define how AI may and may not be used in classrooms and assignments.

2. Instructional Alignment: Ensure tasks still require explanation, reasoning, and transfer.

3. Transparency with Students and Families: Communicate how AI tools operate and how data is used.

4. Equity Review: Evaluate tools for potential bias affecting multilingual learners, students with disabilities, or other student groups.

5. Professional Development: Provide educators with guidance on integrating AI while preserving cognitive ownership.

Leadership Reflection

Artificial intelligence will continue to reshape educational environments. The question for school systems is not whether AI will be used, but how intentionally it will be governed.

When districts adopt AI without clear instructional principles, technology may quietly redefine learning priorities. When leaders evaluate tools through ethical and instructional lenses, AI can expand the capacity of schools while preserving the human work of education.

Leadership Question: Does this technology strengthen the ability of schools to cultivate thinking, judgment, and intellectual ownership? Or does it simply make the production of answers more efficient?

The answer determines whether artificial intelligence becomes a tool for learning – or a substitute for it.

REFERENCES

Bjork, R. A., & Bjork, E. L. (2011). Making things hard on yourself, but in a good way: Creating desirable difficulties to enhance learning. In M. A. Gernsbacher, R. W. Pew, L. M. Hough, & J. R. Pomerantz (Eds.), *Psychology and the real world: Essays illustrating fundamental contributions to society* (pp. 56–64). Worth Publishers.

Bransford, J. D., Brown, A. L., & Cocking, R. R. (Eds.). (2000). *How people learn: Brain, mind, experience, and school*. National Academy Press.

Darling-Hammond, L., Flook, L., Cook-Harvey, C., Barron, B., & Osher, D. (2020). Implications for educational practice of the science of learning and development. *Applied Developmental Science, 24*(2), 97–140. https://doi.org/10.1080/10888691.2018.1537791

Dweck, C. S. (2006). *Mindset: The new psychology of success*. Random House.

Floridi, L., Cowls, J., Beltrametti, M., Chatila, R., Chazerand, P., Dignum, V., Luetge, C., Madelin, R., Pagallo, U., Rossi, F., Schafer, B., Valcke, P., & Vayena, E. (2018).
AI4People—An ethical framework for a good AI society: Opportunities, risks, principles, and recommendations. *Minds and Machines, 28*(4), 689–707. https://doi.org/10.1007/s11023-018-9482-5

Freeman, S., Eddy, S. L., McDonough, M., Smith, M. K., Okoroafor, N., Jordt, H., & Wenderoth, M. P. (2014). Active learning increases student performance in science, engineering, and mathematics. *Proceedings of the National Academy of Sciences, 111*(23), 8410–8415. https://doi.org/10.1073/pnas.1319030111

Hammond, Z. (2015). *Culturally responsive teaching and the brain: Promoting authentic engagement and rigor among culturally and linguistically diverse students*. Corwin.

Hatano, G., & Inagaki, K. (1986). Two courses of expertise. In H. Stevenson, H. Azuma, & K. Hakuta (Eds.), *Child development and education in Japan* (pp. 262–272). W. H. Freeman.

Hattie, J. (2009). *Visible learning: A synthesis of over 800 meta-analyses relating to achievement*. Routledge.

Immordino-Yang, M. H. (2016). *Emotions, learning, and the brain: Exploring the educational implications of affective neuroscience*. W. W. Norton.

Posner, M. I., & Petersen, S. E. (1990). The attention system of the human brain. *Annual Review of Neuroscience, 13*, 25–42. https://doi.org/10.1146/annurev.ne.13.030190.000325

Selwyn, N. (2019). *Should robots replace teachers? AI and the future of education*. Polity Press.

Sweller, J. (1988). Cognitive load during problem solving: Effects on learning. *Cognitive Science, 12*(2), 257–285. https://doi.org/10.1207/s15516709cog1202_4

Made in the USA
Coppell, TX
25 March 2026

74638579R00144